D1032582

Stavros Kromidas

Practical Problem Solving in HPLC

Further Reading

Journal of High Resolution Chromatography
ISSN 0935-6304 (12 issues per year)

Stavros Kromidas

Practical Problem Solving in HPLC

Weinheim · New York · Chichester · Brisbane · Singapore · Toronto

Dr. Stavros Kromidas
NOVIA GmbH
Rosenstraße 16
D-66125 Saarbrücken
Germany

Originally published in German by Hoppenstedt Verlag, Darmstadt
under the title "HPLC Tips"

Translator: Dr. Aran Paulus and Dr. Georg Mozgovoy

Library of Congress Card No.: applied for

A catalogue record for this book is available from the British Library

Die Deutsche Bibliothek – CIP Cataloguing-in-Publication-Data
A catalogue record for this publication is available from Die Deutsche Bibliothek

ISBN 3-527-29842-8

Printed on acid-free and chlorine-free paper.

Composition: TypoDesign Hecker GmbH, D-69181 Leimen
Printing: Strauss Offsetdruck, D-69509 Mörlenbach
Bookbinding: Buchbinderei J. Schäffer, D-67269 Grünstadt
Cover Design: Schulz Grafik-Design, D-67136 Fußgönheim
Printed in the Federal Republic of Germany

Longum iter est per praecepta,
breve est efficax per exempla.

Seneca

Foreword

Nothing demonstrates the importance and maturity of High Performance Liquid Chromatography (HPLC) more than this compendium of practical wisdom about how to master the complexity of instrumentation and the problems associated with the chemical aspects of the technique.

We shall soon celebrate the centennial of the introduction of chromatography by T.M. Tswett, who first demonstrated the concept and practice of differential migration processes which have revolutionized analytical chemistry over the past forty years. In the early fifties, gas chromatography lead the way in exploring the tremendous breadth of chromatography and thus the gas chromatograph has become the paradigm of a new era in analytical chemistry. In the late sixties it was followed by HPLC that has become, and still is, the most versatile separating tool using sophisticated instrumentation and a variety of chromatographic systems. Of course, this stems also from the dual nature of chromatography as being not only a precision microanalytical tool, but also an indispensable process for the preparative/production scale purification of biological substances in particular.

After the introduction of HPLC in the late sixties, the technique experienced a meteoric growth and established itself as the leading analytical tool in the pharmaceutical industry. Since then, HPLC has found wide application in all branches of science and technology. Today the worldwide roles of HPLC instruments and supplies amount to over two billion USD, and the market is still expanding further.

The novice may often find the instrument and the bewildering array of columns and eluents nonplussing. Indeed, the complexity is high, but not so high that at the present its use would require an operator who is a highly trained specialist. The erudite books offer little or no help in getting oriented to finding the right one among a half dozen 1/16“ ferrules that look almost the same, but, if an inappropriate one is used in a given fitting, it will be ruined. Dr. Kromidas's book is a gold mine of useful tips. This practice-oriented book does not fall short of explaining the reasons underlying the problem, and what is just as important, it voices caveat from the consequences of the mistakes one can commit in trying to gain control over the instrument and the separation process.

The advent of HPLC has not only brought us elaborate instrumentation, but has also made reversed-phase chromatography the leading modality of analytical liquid chromatography. An estimated eighty to eighty-five percent of separations are carried out by using alkyl silica stationary phases. In the seventies reversed-phase chromatography set a new direction to HPLC by dwarfing the significance of ion-exchange and normal phase chromatography. As a result, a new generation of chromatographers might think of normal phase as reversed-reversed-phase chromatography. It is gratifying that *Practical Problem Solving in HPLC* pays ample attention to the instrumentation, columns, and operation of reversed-phase chromatography

The forty-five families of tips in this book handsomely cover the present scope of HPLC and besides novices, even a seasoned chromatographer can learn a few tricks from it. The author has laid down the links to developments in HPLC which now move forcefully ahead, for instance, the increasing use of the mass spectrometer as the detector for HPLC. However, many other new problems, as well as opportunities, are coming from the employment of high voltage to bring about separations by capillary electrochromatography and by its cousin, high performance capillary electrophoresis. The new techniques require thorough familiarity with classical HPLC, that stays uncontested the chief method of chromatographic analysis, and inspiration and knowledge to master many of the practical aspects in the future ought to come from books like *Practical Problem Solving in HPLC* . It is concise yet rich in practical information, a combination that would be difficult to find in print elsewhere. It helps everybody to be a better practicing chromatographer and may give relief to many who have difficulties in gaining control over the instrument and the chromatographic process at large.

December 1999

Professor Csaba Horvath
Department of Chemical Engineering
Yale University, New Haven, CT, USA

The Author

Dr. Stavros Kromidas, born in 1954, studied chemistry at the University of the Saar in Saarbrücken, and was awarded his doctorate in 1983 by Professor Engelhardt and professor Halász for his work on chiral phases in HPLC. From 1984 to 1989 he was the North Germany Sales Manager for Waters-Chromatography, Eschborn. Since 1989 he has been Managing Director of NOVIA GmbH, a consultant company for analytical chemistry.

Dr. Kromidas has worked in the area of HPLC since 1978, and since 1984 he has given lectures and refresher training courses. At the beginning of the 1990s, quality improvement in the analytical laboratory became a further work area. This involves optimization of the efficiency of processes in the laboratory from an integrated viewpoint. Dr. Kromidas is the author and co-author of several articles and the following books (in German): Quality in the Analytical Laboratory, 1995, VCH; Validation of Analytical Methods, 1999, Wiley-VCH; Handbook on Validation in Analytical Chemistry, to be published in 2000.

Preface to the English Edition

HPLC Tips, the "yellow book", was a great success in the German-speaking area, and I hope that the English edition will help users all over the world to accelerate their understand ing of HPLC also. In the English version there is some additional information and some more recent results. Because of the importance of the separation of ionic compounds on RP material, the reader will find a chapter by LoBrutto and Kazakevich on this subject. I very much hope that the reader will find some of the hints useful for his or her everyday work. I offer my sincere thanks to Dr. Steffen Pauly of Wiley-VCH who was responsible for the realization of this project.

Saarbrücken, October 1999 Stavros Kromidas

Preface to the German Edition

Our professional daily life confronts us with a multitude of questions. When I was a small boy my grandfather impressed me by always having an answer at his fingertips, no matter what was the question I asked him. His answers were always practical and understandable. His expert knowledge coincided with his experience so that he could describe things clearly in their context.

Since that time, real, solid things have fascinated me – but so also have theories. The present book aims to take account of both.

To reach this goal for a readership with many different backgrounds is not easy. I hope I have been reasonably successful.

The HPLC „Tips" are all about fast answers and help. At the same time they try to point out connections and give explanations in compact form. Language, style and construction of the book serve only one purpose: to make it an easy-to-read companion in the HPLC laboratory. It should not be thought of as a textbook. The reader should acquire the basics of HPLC from the literature on the subject.

I am grateful to my colleague Christine Mladek for the idea of the "General Tips for Newcomers" and for many helpful and intense discussions. A cordial "thank you" also goes to my colleague Anne Weitz-Hartwich for her conceptual inspirations and a critical review of the manuscript, and Mrs Marion Abstiens has prepared a perfectly printable text from this. It was a great pleasure to cooperate with Mr. Rainer Jupe and Mr. Robert Horn of the publishing house in such an informal and very pleasant way.

Saarbrücken, November 1996 — Stavros Kromidas

Contents

1. Introduction

1.1 How to use this book

A short and not too serious look at the name HPLC is followed by a list of frequently used abbreviations and symbols and some tips for newcomers to the subject, including a checklist to be used before doing an HPLC run. Then comes a brief explanation of some important chromatographic expressions. This is intended as a refresher; it cannot replace a study of HPLC theory in an appropriate textbook.
The main part of the book is a series of "Tips", grouped under three headings:

- Simple tests and decision criteria
- Problems and their solution
- Tips for optimization of the separation.

The division of the tips into three topics does not follow hard and fast rules, for the line between "error recognition" and "optimization" is not sharp. Each tip is a complete discussion of a case, and the reader can easily "jump" between the various blocks. Some important facts are discussed in several paragraphs. In the text, additional cross references are given to further tips with related topics. Therefore, all the tips are numbered.

The appendix includes some further information about HPLC.

1.2 HPLC – the development of a name

"Once upon a time, there was a Mr. Tswett, and many years before him a Mr. Runge, who recognized the adsorption characteristics of lime and paper and ...", so the story of chromatography starts. The history of chromatography is probably well known, as well as various anecdotes about the subject. Therefore, we will skip it here.

However, to avoid leaping into the subject with unseemly haste, let us follow the – not too serious – development of the expression *HPLC*.

The beginning of HPLC – **H**igh **P**ressure **L**iquid **C**hromatography – coincides with the culmination of the swinging sixties and the time of the hippies. The remarkable thing about the new technique was the high pressure (to avoid confusion we should speak about high column pressure) that gave the name for this technique.

At the end of the 1970s, the technique was unofficially renamed to **H**igh **P**erformance **L**iquid **C**hromatography, thanks to improved instrumentation and commercially available finely divided stationary phase materials. The instruments were very expensive, and owning one of the **H**igh **P**rice **L**iquid **C**hromatography high-tech machines very often was a point of honor (**H**igh **P**restige **L**iquid **C**hromatography).

The triumphant advance of HPLC began at the beginning of the 1980s. HPLC was in very great demand. From this, a rapid dissemination in the analytical laboratories was a logical consequence. A number of companies were established with different ideas about set-up, user friendliness, important and essential features. From then on everybody was talking about "HPLC" and wanted to get on board. What did people associate with this name?

Users wanted good separations: **H**igh **P**eaks **L**iquid **C**hromatography.

The management of the HPLC companies saw profits: **H**igh **P**rofit **L**iquid **C**hromatography.

Marketing needed more effective advertising: **H**igh **P**ropaganda **L**iquid **C**hromatography.

Last but not least, some sales representatives were not short on promises during their sales pitch : **H**igh **P**romise **L**iquid **C**hromatography.

So, HPLC grew more and more and enjoyed the increasing interest of the analytical community. At meetings, discussions were more heated than debates at the Bundestag. Which is better, low pressure or high pressure, modular or compact instrument design? Is reversed phase an adsorption or a distribution mechanism? And so forth. The quest for high plate numbers and shorter analysis times has outshone Olympic disciplines. The most courageous scientists debated the possibility of 500 000 to 1 000 000 theoretical plates: **H**igh **P**hilosophy **L**iquid **C**hromatography.

And today? There is so little time available today. Time to really work at HPLC, time to look at what HPLC is – a really nice separation method: **H**ighly **P**olite **L**iquid **C**hromatography. Instead of getting it himself, a newcomer will have the equipment installed by the service engineer of the manufacturer, will get a short "introduction" by an experienced, certainly helpful, but stressed colleague: "HPLC is a piece of cake. You push this button on the left to start the instrument, then you move to the table, pick a method, click OK and ... see there is your peak of benzyl noviate." It is clear that unnecessary difficulties arise due to the chronic lack of time in daily life. Then HPLC is blamed for it, becoming a **H**igh **P**roblem **L**iquid **C**hromatography. HPLC might be a scientific discipline – or shall we say a science on its own – but we should meet the strange phenomenon, **H**igh **P**hantasy **L**iquid **C**hromatography, with some composure. Somehow, it always works. Equipped with high initiative, a few important rules and a lot of pragmatism, we will carry the day: **H**igh **P**ragmatic **L**iquid **C**hromatography. I think nevertheless, HPLC gives us a lot of fun, it is our beloved **H**igh **P**leasure **L**iquid **C**hromatography.

1.3 Frequently used abbreviations and symbols in this book

Abbreviations

ACN, MeCN	Acetonitrile
DMSO	Dimethyl sulfoxide
EDTA	Ethylenediamine tetraacetate
Iso-OH	*Iso*-propanol
MeOH	Methanol
LOD	Limit of detection
φ	Phenyl
PIC	Paired ion chromatography (trade name of Waters)
RP	Reversed phase
TEA	Triethylamine
THF	Tetrahydrofuran

Symbols

A	Area
a	Separation factor (earlier: selectivity factor)
Δ	Difference
d_p	Partial diameter
ε_T	Porosity: volume occupied in the column by the stationary phase. Porosity is expressed as a fraction; for example, for RP columns it is approximately 0.7 (70 %).
F	Flow
H (HETP)	Height equivalent of a theoretical plate
I.D.	Internal diameter of the column
k (k')	Retention factor (formerly capacity factor)
L	Length of the column
λ	Wavelength
N	Number of theoretical plates
P	Pressure
pK_α	Isoelectric point; pH at which the concentration of charged and uncharged molecules is identical
R	Resolution
T	Temperature
t_m or t_0	Death time
t_G	Gradient time
t_R	Retention time
$t_{R'}$	Net retention time
u	Linear velocity
V	Volume
V_d	Dead volume of apparatus
V_p	Pore volume
w	Peak width at 4 s (13.4 % of the peak height)
↑	Increase
↓	Decrease

1.4 General tips for newcomers

The first pages of this book are dedicated to users "confronted" with HPLC equipment for the first time. If you are already an experienced HPLC user, you can skip this part.

During the first contact with such equipment, you normally have some help from a friendly colleague and/or the opportunity to study the principles of a chromatographic process in a book. Finally, the manuals of the manufacturer are available. In recent years, several helpful books dealing with error recovery and many applications were published, but there is still hardly a publication for the rookie.

The question now is, how to get started and what to do?

What is HPLC anyway?

HPLC is a fast separation technique. The mixture to be separated is transferred to a column with a solvent or a solvent mixture (*eluent/mobile* phase). The column is a tube, in most cases of stainless steel, filled with the *stationary* phase. The separation happens right there in the column. Under optimal conditions the components to be separated pass through the stationary phase at different rates and leave the column after different times. The components (the solutes) are registered by a detector. This information is passed on to the data evaluation unit and the output is a chromatogram. The number of peaks is equal to the number of separated components in the sample (not necessarily of the components actually present!), and the area is proportional to the amount.

How to become friendly with your HPLC equipment?

You find yourself for the first time in front of your HPLC equipment, consisting at least of an *eluent delivery system (= pump), an injector, a column, a detector* and a *data evaluation system*. If you see several separate devices, you have a *modular equipment*. If you are in front of a large box, you will work with a *compact unit*.

There is also the difference between an *isocratic* and a *gradient system*. These are easy to distinguish. If there is only one inlet tube for the eluent, you have an isocratic device, and, if two or more are present, a gradient system. With a gradient system, two or more solvents are continuously mixed during the separation. This mixing can be performed (a) before the pump by a proportional valve, when we are talking about a *low pressure gradient* in which the mixing happens in the normal pressure or low pressure side of the device *before* the pump. If there is one pump per solvent, the mixing happens (b) *after* the pump on the high pressure side. The mixing takes place in a mixing chamber, where the solvents of both pumps meet. Such a device is a *high pressure gradient*.

Sample introduction is done either with a hand injector or a manual valve (in many cases supplied by Valco or Rheodyne) or with an *autosampler*.

The next device is the column, the heart of the unit, where the separation takes place according to the various separation mechanisms. The column is located – hopefully – in a column oven to guarantee a constant temperature and reproducible results. Columns can be filled with various materials. The stationary phase is selected according to the separation problem you are working on. Probably, you will work with a C_{18} column. The stationary phase in this case is a chemically modified silica gel (see below).

The *detector* is most often a UV detector, sometimes a diode array detector (DAD or PDA photodiode array). If you find a different detector in your device, e.g. a fluorescence detector, you can assume you will be working with something special.

For *data evaluation*, a computer with the corresponding software is usually installed. An integrator would belong to an older generation of data evaluation systems. If you are working on a *qualitative* analysis, you "only" have to separate all peaks, e.g. solutes, contained in your sample. If you have to run a *quantitative* analysis, the exact concentration or amount of each component present in the sample has to be determined using standards. The data evaluation is most often done using peak areas, very rarely peak heights (unfortunately ...). In addition to the data evaluation, the computer very often controls the whole device, starting with the pumps, the autosampler, the detector and potentially other peripheral modules.

The operation of the equipment is best explained by a colleague, or when you participate in a seminar organized by the manufacturer. Just make sure that "your" equipment will be included in the practical training sessions.

Before you can start with the first measurement, you must carry out a few general tasks. The equipment should be placed in such a way that it will be easily accessible from both sides, the front and the back. The electrical connections between the single modules should be marked at each end, so that a later rearrangement can be done easily. Do not change those connections for the moment! You should keep an eye on all the electrical connections in case they become loose, thereby causing a bad contact or a total power failure.

The mobile phase is transferred from one module to the next in *capillaries* composed of stainless steel or PEEK (polyetheretherketone). The internal diameter (I.D.) of the capillaries between pump and injector should be 0.5–1 mm, and after the outlet of the injector less than 0.2 mm. Some detectors, such as fluorescence detectors, need a certain back pressure for good operation. This can be achieved with a 0.1–0.2 mm I.D. capillary installed behind the detector. Air bubbles, if present in the system, will then remain in the eluent and not disturb the chromatogram with air spikes. These are *restrictor capillaries,* sometimes also simply called *restrictors.*

All interconnection pieces, ferrules and fittings should come from one manufacturer, because different brand names may have small differences in their dimensions, leading to a small dead volume and consequently to a deterioration of the anticipated separation (see Tip No. 9). In general, connection fittings should be tightened by "feel", since otherwise the screw thread can break off and, in accordance with Murphy's law, will invariably get stuck inside the detector. If you need to apply force to tighten a leak, please make it gentle force!

Now you can get started (or can't you?)

First, you have to ensure that your system is clean. A couple of questions: did somebody use the HPLC equipment before you? If yes, which mobile phase was used? Is the column still in the system or has it been removed?

If you do not know what happened to the equipment before you got to work on it, you should flush it (without the column) at a flow rate of 1 ml/min with a 50/50 mixture isopropanol/water for about 10 min. You should also inject the mobile phase a few times in order to ensure that the old eluent or impurities are removed from the sample injection system. Now you can bring the mobile phase recommended for your method into the system. Again, do not forget the injection system.

In the following, the most important HPLC activities are described in more detail, just in case you do not have a description at hand. Otherwise, make the following assumption: if you have to follow an existing method (SOP, System Operation Procedure), stick to this method!!! Your creativity and experimental skills are an invaluable asset in the HPLC laboratory, but please, at the right time.

Which column do I have to install in the HPLC instrument?

The method description will certainly state the column you should use for your work. If not, refer to Tip No. 2. The most popular column material (stationary phase) is a C_{18}-modified silica gel. This stationary phase and the corresponding mobile phase most often consist of mixtures of water with methanol or acetonitrile, and we are then dealing with *reversed-phase* chromatography. The mobile phase can also contain additives or buffers.

If you have to use non-modified silica gel as column material due to your sample, you are working under *normal phase* conditions, although these are used in only 5–10 % of all routine methods. The most important solvents are hexane or heptane in corresponding mixtures. Referring to other separation mechanisms, only some names are mentioned here: ion exchange chromatography, affinity chromatography, exclusion chromatography, chiral chromatography.

How do I prepare a mobile phase?

Your operating procedure tells you which mobile phase you will need, as well as which chemicals and highly purified solvents you should use to prepare it. Most solvents are labeled HPLC grade and are commercially available from a number of companies.

In HPLC, several mobile phases are used to influence the strength of the interaction between sample and stationary phase. The greater the *elution strength* of the mobile phase, the earlier are the components of the sample eluted. In reversed-phase chromatography, the elution strength increases from water to methanol to acetonitrile to THF.

The mobile phase should always be prepared in the same manner. If your method description does not state exactly how to prepare the mobile phase, use the following sequence when preparing it:

Example: buffer/organic as eluent

- prepare the buffer (p.a. quality) in the desired concentration (do not fill up the measuring flask)

- adjust or measure the pH value (attention: only between pH 2 and pH 8; higher pH leads to dissolution of the stationary phase base material; in strong acids the bonds between the base material and the C_{18} chains break, see also Tip No. 7)
- fill up the desired volume of buffer into a measuring flask
- measure the methanol or acetonitrile in another flask
- finally mix.

This procedure guarantees a reproducible preparation of your mobile phase and avoids problems associated with volume contraction. Always prepare a sufficient amount of mobile phase (approx. 1 l) and degas it. *Degassing* is possible with helium or the built-in degasser (see Tip No. 5). If you use buffers, you should filter them through a membrane. The eluent container should be well covered in order to avoid dirt contamination.

The first sample

After having prepared your equipment in the described way, you can attach the source of prepared mobile phase. Now you should leave the equipment for a little time to *equilibrate*. This way, manufacturing-induced impurities are flushed out of the column as well as other dirt. During this time, you can prepare your sample. If you are in luck, you will only have to dissolve it in the mobile phase. If not, follow the method described in the operating procedure. All particles should be removed, most simply by membrane filtration. Never forget to test to ensure that your dissolved sample does not precipitate in the mobile phase. Should this happen in your equipment, you will be busy for some time with cleaning or even replacing expensive parts.

Your system is now equilibrated. The time required is somewhere between a few minutes (simple analysis, e.g. a UV detection) and a few hours (trace analysis, e.g. an electrochemical detector). In order to test whether the whole equipment is functioning, inject a *standard mixture*, which is normally specified in the operating procedure. If not, use a mixture of nitromethane, chrysene, perylene, column: C_{18}, mobile phase: methanol. Take a look at the chromatogram. Is the baseline stable with no drift and are the peaks symmetrical? Is the chromatogram after the second injection identical to the first one? If yes, your total system is OK.

But now, let's get going

According to the operating procedure, inject samples for comparison, samples and standards in a predetermined *sequence* and evaluate the resulting chromatograms.

Take your time if your method contains a gradient. A new run should start after 5–10 min at the earliest to ensure that your system is at equilibrium for each injection (see Tip No. 20)

If you have to install a new reversed-phase column, flush it with methanol or acetonitrile before the first run. Take care to keep your equipment free of buffers. Even better, flush your equipment first with an isopropanol/water mixture and then with methanol. Use the same procedure if you wish to go back to your original conditions.

Quitting your HPLC equipment

To finish, a few pieces of advice for the correct shut down procedure.

1. If you know that you will continue working the next day, it makes sense to shut down all instruments except the pump. Keep the computer running. Leave the pump operating at a low flow rate of 0.1–0.3 ml/min. Make sure there is enough mobile phase to avoid running the equipment dry, or even better *recycle* your mobile phase by running the outlet capillary from the detector to the eluent container. The next morning, you only have to adjust your flow and get started.
2. If you want to shut down your equipment for a longer time, flush the buffer out of the whole unit using water. Then flush with 20–30 ml methanol or acetonitrile. Now you can remove the column, close it with end fittings to avoid drying out and store it with solvent for a longer period (acetonitrile is stable for a longer period than methanol because of the hydrolysis properties of the latter).

To have the essential information at a glance, a check list "What to pay attention to before starting a method" and a flow scheme "How to start working with HPLC equipment" follow on the next page. Maybe in your particular case additional or other steps will be necessary. Fill in those steps in both schemes or modify them. Develop your own working documentation that will satisfy *you* and make you feel safe. After a short while, all these steps will be obvious to you. If you have gained practical experience, you can simplify or shorten one or more steps, but remember the rule from real laboratory life, valid in all routine work: Do the same things in the same sequence and you will get comparable results – even if they are wrong!

1.5 Check list for reversed-phase HPLC

What do I have to pay attention to before starting a measurement?

Electrical connections
- Nothing loose?

Capillaries
- Leaks?

Tubes and solvent container
- Remove air bubbles from inlet tubes by sucking solvent with syringe with purge valve open (purge, prime).
- Use covered solvent container to avoid objects falling into it and to minimize evaporation.

Pumps
- Switch on pump and look at waste container. Does it drip into container? If not, check if mobile phase flows through inlet tube. Most frequent cause for missing flow: air in pump.
  Does it leak or is it wet (touch the seals)? Do you see crystals when using buffered mobile phases? Is there anything unusual about the pump noise?

Mobile phase
- Always prepare mobile phases with HPLC-grade solvents.
- Prepare a sufficient amount
- Buffer concentrations between 20 and 100 mM (see Tip No. 8).
- For buffered mobile phases, always use membrane filtration, degas with helium or with degasser.
- If possible, avoid adding aggressive components such as trichloroacetic acid to the mobile phase.

Injector
- With manual injector, make sure there is a container under the overflow, keep injection needles clean to avoid contamination, flush with isopropanol if necessary.
- For some autosamplers, washing solution must be connected. For reversed-phase separation, add 10–20 % methanol to the water to avoid micro-organism growth.

Column
- Always use the same equilibration procedure.

Detector
- If you are working with UV detectors, check lamp energy.

Waste
- Use sufficiently large container.

Data evaluation
- Are the preset integration parameters and sample rates all right?

Flow scheme for RP methods:
How do I start working with the HPLC equipment?

First check the condition of the system.
Is the mobile phase for your method already equilibrated? Do you have to re-equilibrate for your mobile phase? Or have you no knowledge about this?

Flush your equipment with water/methanol or water/acetonitrile mixture, roughly 10 ml of each, with the column removed. This can do no harm. If you are working with an organic solvent, an intermediate flushing step with methanol and methylene chloride is necessary before switching to your mobile phase conditions. Attention: with buffered mobile phases never switch directly to 100 % acetonitrile or vice versa! Otherwise the buffer salt will precipitate resulting in clogging the equipment.

If you have to re-equilibrate the equipment, first flush out the current mobile phase. For buffer, use water, otherwise water/methanol or water/acetonitrile mixture.

Install the necessary column.
If you are using a new column, condition it according to the instructions of your supplier. If your mobile phase contains buffer, again flush first water/methanol or water/acetonitrile mixture through system, then mobile phase.

If your system is ready for use, go on to the next step.
Prepare the necessary mobile phase according to the method description, membrane filter buffer, degas, connect to inlet tube, if necessary remove air bubbles with syringe or purge.

Check if there is a waste receiver after the detector. Check injector; see if waste receiver is under the overflow. Using an autosampler connect wash solution if necessary.

Switch instruments on in the sequence pump, injector, detector, data evaluation (except the case your PC controls your pump).

Increase flow in 0.2 ml/min steps until desired flow rate is reached (take a look at the back pressure!)

Take your time, so the column can equilibrate (always wait the same time!). In the meantime, prepare samples. Check for precipitation when mixing mobile phase and dissolved sample. If you have precipitation, try a different solvent.

If everything is OK, inject standard. Check if the obtained chromatogram superimposes with a reference chromatogram; are the peak data (area, height, asymmetry) and retention times unchanged? If so, your system is ready and you can start your measurements.

If you quit for the day but continue the next day, set mobile phase recirculation flow rate to 0.2 ml/min. Switch off all modules except the pump.

If you are quitting for a longer period of time, flush the mobile phase out of the system. If you have used buffer, first flush with water, then methanol or acetonitrile, each about 20–30 ml. Remove column, use end fittings to avoid drying and record the conditions in a column log book.

Switch off instruments in the sequence data evaluation, detector, injector, pump. To switch off pump, decrease flow rate in 0.2 ml/min steps.

1.6 Some important chromatographic terms

Symbols/names/formulas for the calculation	What does this mean?	What can I do with it?
t_o or t_m Dead time, very often in the lab. jargon: front $$t_0 = \frac{L \cdot q \cdot \varepsilon_r}{F} = \frac{L}{u}$$	Retention time of a solvent molecule or an inert substance (no interaction with the stationary phase) in the system: that is the time from the injection of a non-retarded substance to the appearance of a peak in the detector (apex of the peak).	Change of t_m means (a) either change of flow rate (pump, leakage, see Tip No. 26) or (b) change of the column dimensions or the packing density.
t_R Retention time $t_R = t_m + t'_R$	Retention time of a retarded substance in the system. That is the retention time in the mobile phase t_m, plus the retention time in the stationary phase t'_R	At otherwise constant conditions (see below), possibility of comparing the behavior of substances in similar or in different systems, e.g. comparison of two columns. Requirement for – equal column dimensions – equal flow – equal packing density However, the *k* value is the more robust criterion for the comparison (see below).
t_z Interparticle time $$tz = t_0 - \frac{V_p}{F}$$	Retention time before t_m: this component cannot penetrate into the pores because it is too large or it is a strong ion – it is excluded (see Chapter 5: Retention of ionizable components in reversed-phase HPLC)	Do I have the right chromatographic system for these substances?
k' Capacity factor, according the new IUPAC terminology: "*k*", retention factor. $$k = \frac{t_R - t_0}{t_0}$$	Rate representing the affinity of this substance for the stationary phase in this chromatographic system (chromatographic system: stationary phase, eluent, temperature). How much longer does this substance remain in/at the stationary phase in comparison to the mobile phase?	1. Using the comparison of *k* values at equal chromatographic conditions I can always compare results directly, and of course also if the flow, the packed density or the inside diameter of the column is different in two cases! This is possible because the *k* value is independent of the flow and the dimensions of the column. 2. Reference for the next step at the optimization. $k < 1 \rightarrow$ substance comes too early. $k \approx 2–5 \rightarrow$ (optimal area), see Tip No. 41.
α Selectivity factor, according the new IUPAC terminology: separation factor $$\alpha = \frac{t'_{R2}}{t'_{R1}} = \frac{k_2}{k_1}$$	Rate representing the selectivity; i.e. for the separation capability of a chromatographic system for certain substances. Relationship of the retention time of two substances in the stationary phase. a > 1 is *the* fundamental requirement for a separation in chromatographic techniques.	Increasing the selectivity is very often the most elegant but very often a difficult method to improve the resolution (see below).

Symbols/names/formulas for the calculation*H*	What does this mean?	What can I do with it?
H Theoretical plate height (height equivalent of a theoretical plate) $H = \frac{L}{16} \cdot \frac{w^2}{t_R^2}$ and *N* theoretical plates number (number of theoretical plates) $N = \frac{L}{H} = \frac{16\, t_R^2}{w^2}$	Rate for the band broadening (peak broadening) of a substance in HPLC equipment. The smaller the *H* value, the bigger the plate number. This means that the better the column is packed and the smaller the dead volume of the instrument, the sharper the peaks will be: the efficiency is good.	Objective criterion at the comparison of the packing of two columns. For example, a column with 10 000 plate number gives smaller peaks than one with 5000. The selectivity used is decided by whether a separation is really to be expected! Be aware if you compare *N* values that the number is influenced by the viscosity of the eluent, the injection volume, the retention time, the flow and the temperature.
R Resolution $R = \frac{1}{4}\sqrt{N}\,\frac{(\alpha - 1)}{\alpha} \cdot \frac{(k)}{(k+1)}$	The degree to which one peak is separated from another. Distance between peaks at the peak bases.	This is *the* separation criterion for a chromatographic system, since "everything" depends on the resolution, which influences the separation: capacity, selectivity and efficiency.
V_d Dead volume A symbol for the dead volume; very rarely used. $V = F\ t$	Volume of the equipment from the injection to the detector cell – without the column! (See Tip No. 9) Important for isocratic equipment. Sometimes one says "dead volume" and means the above-described volume including the column.	The smaller the dead volume, the sharper are the peaks. Rules of thumb: Ca. 20–25 µl very good, ca. 30–60 µl good enough for a 4 mm column; considerable tailing at 2 mm columns and/or 3 µm material. From ca. 70 µl unnecessary band broadening, above all at the early peaks.

Summary

k **Capacity**: rate for the interaction of a given substance in a chromatographic system. This is the rate that indicates how much longer a substance remains in the system than a substance which does not interact with the stationary phase.

α **Selectivity**: rate for the separation capability of a chromatographic system for two or more substances; ratio of the retention times of the two substances at the stationary phase. *k* and *a* are influenced only by the "chemistry"; i.e. temperature, stationary and mobile phase, pH, ionic strength, additives in the mobile phase. For the isocratic mode, capacity and selectivity are independent of equipment design as well as flow, packing density and column dimensions. (In reality the dead volume influences the *k* value very little, but let us be a little generous.)

N **Efficiency:** rate for the band broadening of a substance in the system; do I get sharp or broad peaks?
For an inert substance (elution at t_m) only the "physics" plays a role; i.e. diffusion coefficient, viscosity, linear velocity (mm/s), dead volume of the device, particle size, quality of packing, column length. At an actual separation also the "chemistry" naturally is important, because the kinetics of the adsorption ⇔ desorption, for example, depends on the surface of the stationary phase and the temperature.

R **Resolution**: distance between the peaks, which is really what interests the "normal" user. The resolution is influenced by the above three factors, which again means that a separation optimization can be reached exclusively through change of *k*, *a* and *N*. The most effective way to reach an improvement of the resolution is through a change of the selectivity (see Tips Nos. 40 and 41).

2. Simple Tests and Decision Criteria

Tip No. 01 What does the name of a column material tell us?

The Case

Let us look at names of computers: 200 MHz tells us the clock speed of the computer; 32 MB RAM says something about the random access memory; 4 GB describes the size of the hard disk. Visio, Presario and others are smart marketing names that do not tell us anything about the characteristics of the product. The situation is similar for HPLC stationary phases. What do the manufacturers tell us in a name of their product?

The Solution

There are stationary-phase material names that give no information, such as INTERCHROM and ASAHIPAK, and others that do tell you something about themselves, such as LiChrospher and Inertsil. This subject cannot be dealt with in detail, but in the Table below you should find some help. Listed are numbers, letters, prefixes and suffixes from names that give an indication of properties of column materials.

Table 1-1. Names of HPLC stationary-phase materials.

From the names ...	... the information
Nature of the stationary phase	
-Sil, Si, Silica, -spher, -sorb	**Sil**ica gel
-Alox, A, Alumina	Aluminum oxide
-HP, HA	**H**ydroxy**ap**atite
-CEL	**Cel**lulose
-GEL	Polymer
-Silica "B"	Silica gel manufactured by a new procedure and, in comparison to a silica "A", without metal ion contamination
AQ, AQUA	**Aq**ueous: phases with hydrophilic endcapping for the separation of polar, organic analytes with hydrophilic eluent or pure water (polar RP phases).
Physical characteristics of the stationary phase	
-spher	Spherical material
-sorb	Irregular, broken material Exception: Spherisorb: spherical
-WP	**W**ide **p**ore, e. g. 300 Å pore size
NPB, NPR, NPS	**N**on-**P**orous **B**eads, **N**on-**P**orous **R**esins, **N**on-**P**orous **S**ilica: suitable for quick separation (it is necessary to have a device with low "death" volume)

From the names ...	... the information
EPS	**E**xtended **P**olar **S**electivity: C_{18} with polar characteristics (see also AQ)
HD	**H**igh **D**ensity: phase with high carbon content, ca. 20 % C, stable
CP	**C**apsulate **P**olymer: surface-coated polymer
100, 120, 300, 1000, 4000	Pore diameter (a large number) of the stationary phase, important for the separation of large solutes, e.g. proteins.
3, 5, 10	Particle size (a small number) e.g. 5 µm is a medium particle diameter
Modified stationary phases	
RP_x	Reversed phase, x = 2, 8, 18
OD(S), C_{18}, RP_{18}, 18	**O**cta**d**ecyl**s**ilane: C_{18} alkyl chain
ODS I, II, III (or 1, 2, 3)	Similar to computers – the new generation of stationary phase. Unfortunately there is no system in the naming. Sometimes, "II" is "endcapped" (i.e. second silanization, see Tip No. 3) and "I" is not, sometimes "II" is endcapped better than "I", sometimes it is double endcapped and sometimes the procedure for endcapping has been optimized.
OS, $C_{8,}$ MOS	**O**cta**s**ilane, C_8- alkyl chain, **M**ethyl **O**ctyl **S**ilane
APS, NH, NH_2, Amino	**A**mino **P**ropyl **S**ilica, modified with a $(CH_2)_3NH_2$ group
PH, φ	**Ph**enyl group
e, E	**E**ndcapped, second silanization of RP stationary phases
ES	**E**ndcapped **S**ilica
NE	**N**on-**E**ndcapped
Suitability for a specific separation problem	
– **Sugar**pac	for sugar separations
– TSKgel **DNA**-NPR	for DNA fragments and nucleic acids
– **Pep**MAP C_{18}	for peptides
The following letters suggest ion exchange columns	
AX, SAX, WAX, SCX, SC, CX, IEC, IEX	**S**trong **A**nion E**x**changer, **W**eak **C**ation **E**xchanger, **I**on **E**xchange **C**hromatography, **I**on **Ex**changer, etc.
Names of specially treated stationary phases: suitable for the separation of basic solutes	
Inertsil	"Inert towards bases"
Symmetry	"Delivers symmetric peaks"
Select B	"Separates bases"
deactivated phase	Homogeneous surface, inactive OH groups
-pur, purospher	Very pure stationary phase, again no metal ions
"shield"	Protective group on alkyl chain to protect silanol groups
SB	**S**table **B**ond
SP	**S**terically **P**rotected
AB	**A**cids and **B**ases (suitable for separation of acids and bases)
ABZ plus	**A**cids, **B**ases and "**Z**witter" ions: (suitable for separation of acids, bases and "Zwitter" ions; introduces polar "protective group")

From the names ...	... the information
BDS	**B**ase **D**eactivated **S**ilica, (not activated = suitable for bases)
SilicaROD	The stationary phase is constituted as a solid rod; no classical particles („monolithic phases”).

Conclusion

There is no conclusion. If you really want to know more about your stationary phase, call your supplier.

Assuming your sales representative knows his stuff, try to use your charm and wit, involve him in a scientific discussion or explain to him how much he already makes or will make from your orders. Good luck!

Tip No. 02 Is this C_{18} column the right choice for my sample?

The Case

In view of the considerable number of C_{18} columns available, it is unreasonable to test by trial and error many stationary phases for a new problem. If you do not have more detailed information, you should try to limit your choice. There are a few rules.

The Solution

The following solute characteristics are helpful for a first limitation:

- solubility
- chemical nature (pH-value)
- molecular weight

For a start, you should decide whether your solute can in principle be separated on a C_{18} silica gel based stationary phase in a classical reversed-phase system. Subsequently, you can eliminate some columns based on solute characteristics and can focus on others.

Table 2-1. Solute characteristics and column choice.

Solute characteristics	What does this tell me?
Solubility	
The sample is *only* soluble in hexane/heptane.	The sample is most certainly very nonpolar, elution on a C_{18} column will most likely be very difficult. Alternative choice: SiO_2, CN or NH_2.
The sample is unstable in water.	Use non-aqueous chromatography (NAC): C_{18} and acetonitrile/THF/methanol mobile phase.
The sample is soluble in an acidic or alkaline solvent and most likely can be analyzed in it.	Use a suitable C_{18} column, see below.
Chemical nature	
The solutes possess a similar polarity, there are only minor structural differences.	A C_{18} column is in this case rather non-selective (perhaps polar C_{18}-phases suitable); better alternative: SiO_2, CN, NH_2, microcellulose.
All solutes are present as ions.	C_{18} and ion pair reagents[1)] in the mobile phase or ion exchange column are the two alternatives.
The sample contains acids (measure pH).	Most modern columns have problems with acids. Use "old" classical ones like Hypersil ODS, Resolve or Nucleosil 100. Use an acidic buffer as eluent. (see Tip No. 44, No. 45).

1) Ion pair reagents, "PIC" reagents (PIC is a trade name of Waters) are added to the eluent if strong acids or bases have to be separated on RP columns. PIC-A reagents are suitable for the separation of acids, e.g. dibutyl- or tetrabutylammonium phosphate or chloride (pH ≈ 7.5). PIC-B reagents are suitable for the separation of bases: B_5 to B_{10}, penta- to decasulfonic acid (pH ≈ 3.5).

Solute characteristics	What does this tell me?
The sample contains bases.	Do not take not endcapped phases like Hypersil ODS, Resolve oder LiChrospher. If you are obliged to do this (because of SOPs), you should add an amine to the eluent (see Tip No. 28 and 45).
The sample contains very polar, but nonionic solutes	If you would like to use a C_{18} column, choose one with a high carbon content (above 18 %), in order to achieve a long column lifetime with the highly aqueous mobile phases. Maybe the better alternative: C_4, CN, diol, phenyl
Molecular Weight	
(a) ca. 250 < MW solutes < 400	It is possible that you will observe exclusion effects on an Si60 stationary phase with this relatively high molecular weight; some peaks may elute before t_m (see Tip No. 41) because the pore diameter of Si60 is "only" 60 Å. Maybe you would like to use both mechanisms (exclusion and adsorption) to increase your selectivity. Then an Si60-C_{18} column is ideal. A good example is the separation of substituted phenols.
(b) ca. 400< MW solutes < 800	In an isocratic run, larger solutes elute slowly from a C_{18} surface; gradient conditions pose no problems. Alternatives for isocratic runs: C_8, C_4, C_1, nonporous materials (NPS)
(c) ca. 800 < MW solutes < 2000	Gradient runs are with short columns possible; alternatively, exclusion chromatography

Conclusion

These rules should be considered as the alternative approach if you have no information about your stationary phases and no way to get it. Often it is easier to insist on help from your (internal) customer or to quiz your supplier for the most suitable column or to run a detailed literature search. Think also of the opportunities of the internet!

Tip No. **03**

Why are polar solutes well separated with one C_{18} column and hardly at all with another?

The Case

You would like to separate polar solutes with reversed-phase chromatography. Let us assume the compounds are basic and you would like to test several columns for suitability. Of course, you do everything just right and choose "good" columns with an endcapped stationary phase[2)]. Despite your careful choice, you obtain under identical chromatographic conditions (mobile phase composition, pH, temperature, etc.) reasonable results with one column, whereas the use of a second column, endcapped as the first, results in tailing peaks. Why?

The Solution

The surfaces of several silica gels are partly contaminated with metal ions, depending on the different manufacturing procedures. On the other hand, there are big differences between materials according to the concentration of "active" silanol groups. The situation is quite complicated, but the above-mentioned two factors are the reasons in simplified form. For the acidic or basic pH of silica gels see Table 3.1

Table 3-1. pH values of proprietery silica gels.

Silica gel	Batch	pH
Superspher Si 60	L293786	3.3
LiCrospher Si 60	L165018	3.3
Superspher Si 100	L222511	3.5
LiCrospher Si 100	L321017	3.5
TSK-Gel	A3601	3.5–4.0
Zorbax BP-Sil	8341-114	3.9
Purospher	FE109537	4.0
Reprosil Pur	80101	4.0
Gromsil CP	A0201	4.4
Zorbax RS-Sil	G12840	4.4
Separon	SI VSK	4.5
Reprosil AQ	20405	5.0
Novapak	75A	5.1
LiChrospher Si 100	YE 187	5.3
Porasil		5.4
LiChrospher Si 300	V V 1879	5.5
Nucleosil AB	8011	5.5
YMC Pro/AQ	80182903/4	5.5
Megapharm	AN1906135	5.5
Nucleosil HD	8021	5.6

2) Notice: Endcapped C_{18} phases are suitable for the separation of basic solutes. The C_{18} phase has been treated in a second step with a small silane molecule. Because it is small and therefore sterically compatible, it can reach the deeperlying, less reactive silanol groups. As a result you will get a surface that contains less active silanol groups. Fewer problems therefore arise during the separation of polar groups, because no additional interactions of a polar nature are possible. Endcapping can be found sometimes within the manufacturer's information about the column marked as "e" or "E", see also Tip No. 1.

Silica gel	Batch	pH
Nucleosil 100	9081	5.7
Nucleosil Protect	7035	5.8
Nucleosil 100	8071	5.9
Prontosil	970127BQP	5.9
SMT		6.0
Purospher Star		6.0
SMT		6.0
Spherisorb	18A	6.1
Platinum EPS/C18	26/146/26/189/5	6.2
LiChrospher Si 4000	YE 248	6.2
Ultrasep		6.2
Nucleosil 50	7013	6.3
LiChrosorb Si 60/Si 100	L309019	6.5
LiChrosorb Si 100	F 1987	7.0
Kromasil	AT0104	7.0
Porasil	5/16/80	7.2
Partisil 10	A 246/2	7.5
Polygosil 60-5	59	8.0
LiChrosorb Si 60	YE 93	8.1
Spherosil XOA	400	8.1
Hypersil	80-5-13	8.1
Gromsil ODS	2605	8.7
LiChrospher Si 500	YE 209	8.8
Hypersil	1/717	9.0
LiChrospher Si 1000	YE 338	9.2
Spherisorb S 10 W	764	9.5
LiChrospher Si 500	YE 52	9.9

There is a big range, is there not?

The separation of polar solutes is of course influenced by the pH of the stationary phase. The additional polar interaction and the slow kinetics result in a "chemical tailing". Figure 3-1 shows the separation of phenols (acidic solutes) on a basic stationary phase (Hypersil) with a lot of active silanol groups and on an acidic stationary phase (Zorbax). Although the separation is obtained in a normal phase mode with a nonpolar solvent, Hypersil is able to form ionic interactions in addition to the expected van der Waals interactions, resulting in strong tailing. With Zorbax, tailing is less pronounced because an ionic interaction is not as likely.

Conclusion

Since the physical characteristics of silica gels remain unchanged during the chemical modification to C_{18} phases, you can make a rational choice of the column material for method development of polar solutes. But again, the pH of the material can only be *one* of the criteria for choosing the right column.

The good news:

Luckily, the pH of silica gels is a constant characteristic of the material. Evaluations in our laboratories on sample batches show that the pH of certain silica gels delivered in 1983, 1988, 1996 and 1999 only deviated by less than 0.1 pH units.

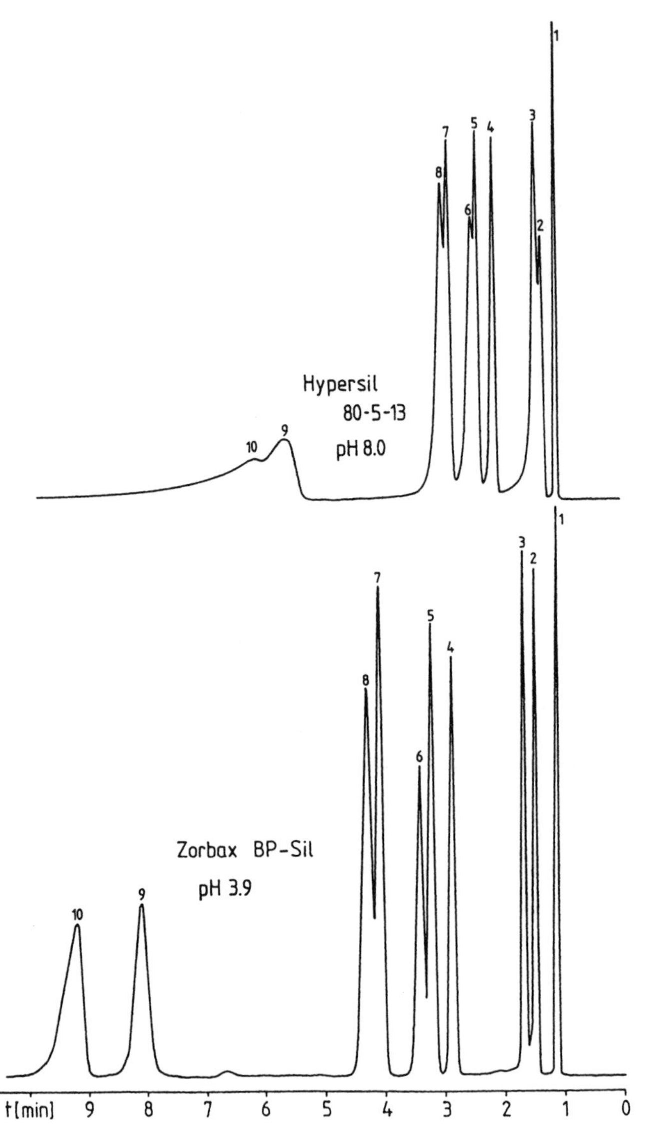

Figure 3-1. Separation of phenols on two silica gels (source: Prof. Engelhardt).

Tip No.
04

How can I clean the RP phase quickly?

The Case

Hydrophobic organic molecules such as lipids and large organic molecules readily stick to RP phases, especially to C_{18} materials. Occasionally this results in a high back pressure and almost always a decrease in separation performance. Broad and tailing peaks can be observed; sometimes "ghost peaks" are detected. The appropriate solution would be rinsing the column with methanol or acetonitrile. However, this procedure is time-consuming and laborious, and often deadlines are pressing. What to do to save time?

The solution

Inject 100 µl or 200 µl methanol or acetonitrile. (Check before the injection that acetonitrile does not cause a precipitation of buffer-containing eluents!) If organic solutes are absorbed on the reversed-phase surface, you should get a large peak. You have performed a kind of displacement analysis with a miniaturized rising step. Repeat this procedure twice or three times. If you still get a "garbage" peak, you have to do the "normal" flush with methanol or acetonitrile. However, very often this procedure works out just fine and you will have the familiar small solvent peak at the beginning of your chromatogram at t_m after just one or two injections. The reversed-phase surface is again clean. Additionally, even stronger eluents such as THF, DMSO or heptane are available; however, their elution strength is really strong. This sounds like a warning, and to some extent it should be. The surface is really effectively cleaned by these solvents, but the selectivity may possibly be altered.

Conclusion

If you assume organic impurities on the surface of the reversed-phase column, first inject acetonitrile. It acts as an indicator of the presence of dirt and flushes out organic impurities. With a little luck (it does happen in HPLC...) the column will be clean. This procedure is certainly faster than the usual rinsing procedure or a change to a new column with the necessary pre-equilibration.

Tip No. **05**

How best do I degas my mobile phase?

The Case

Noisy or drifting baselines and pressure fluctuations are signs of insufficient degassing. You should degas your mobile phase or improve your degassing procedure.

When is this necessary?

Very important	Advisable
Methanol/water as eluent	Acetonitrile/water as eluent
Low pressure gradients	High pressure gradients
Fluorescence, electrochemical and RI detection	UV detection
< ca. 210 nm	> ca. 210 nm

The Solution

In the following two figures, you can see the efficacy of the four most important degassing methods for a polar (methanol, Figure 5-1 left) and a non-polar (hexane, Figure 5-1 right) mobile phase system.

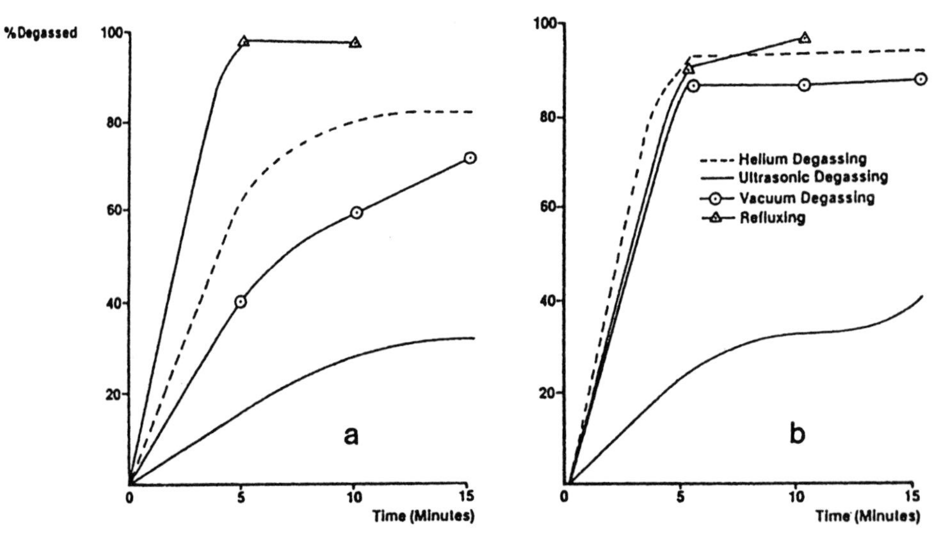

Figure 5-1. Efficacy of degassing methods for a polar (MeOH) and a non-polar (hexane) solvent. (Source: J W Dolan, L R Snyder, Troubleshooting LC Systems, Human Press, 1989)

Some comments:

- *Refluxing* is good, but not practicable
- *Vacuum degassing* is a good method. It is often combined in a one-step procedure with the filtration of buffered solutions. You may use a commercial vacuum degasser, implemented in your pump delivery system.
- *Helium degassing* is also a good method.
- *Ultrasonic degassing* is rather ineffective and only applicable for acetonitrile/water mixtures or if you have a really good pump.

Conclusion

Although some users complain about stability problems of the membranes in commercial degassers from some suppliers, the general experience is positive. The only disadvantage of helium is its price. You should degas thoroughly at the beginning for 10–15 min before either shutting off the helium flow for the rest of the run or, if necessary, see symptoms above, maintain a low helium flow (e.g. 0.5 ml/min).

Tip No.
06

Methanol or acetonitrile?

The Case

Acetonitrile/water and methanol/water mixtures are the most common mobile phase compositions in reversed-phase HPLC. Is there a preference? Which is better: acetonitrile or methanol?

The Solution

Naturally, there is no general answer to this question. Selectivity differences could play a vital role for your separation. For example, some solutes can form methanolates, while acetonitrile stabilizes octahedral Cu^{2+}, Cd^{2+} and Zn^{2+} complexes. Some general differences between the two solvents are listed below:

Table 6-1. Comparison of acetonitrile and methanol.

Positive features of acetonitrile	Advantages
• Low viscosity	
– better kinetics	Sharper peaks
– acetonitrile/water mixtures have lower back pressure in comparison to methanol /water mixtures	Less wear and tear on seals and columns
• Higher elution strength	Lower solvent consumption: you will have the same elution strength at a lower percentage of acetonitrile compared to methanol Silica gel is less prone to dissolve in acetonitrile-containing eluents compared to methanol-containing mobile phases (acetonitrile is less polar than methanol)
• Low solubility of air	Fewer problems with air, less effort for effective degassing
• Low UV absorptivity	Better for detection at 195–200 nm
• Only small pH deviations in aqueous solutions	Better reproducibility especially for separations of ionic solutes
• Acetonitrile is a better solvating agent	Advantages in ion chromatography
• More chemically different from water than methanol	Selectivity differences easier to obtain
• More toxic than methanol	Microbiological growth in the equipment is hindered
• Better for the separation of bases at low pH	Sharp peaks
Positive features of methanol	**Advantages**
• Odor less inconvenient	Better working conditions
• Less toxic	Healthier working conditions
• Better solubility for salts	Danger of precipitation low, even at 100 % methanol in gradient conditions

Positive features of acetonitrile	Advantages
• Methanol/water mixture brings the seals faster into its swelling conditions	The equipment gets faster to working conditions
• In older batches of acetonitrile, impurities (propionitrile, methacrylonitrile) can give rise to "ghost" peaks, problem less known for methanol	Longer shelf live of methanol
• Better for the separation of bases at alkaline pH	
• Lower baseline noise above 220 to 230 nm in methanol/water mixtures	

Conclusion

Obviously, the advantages of acetonitrile outweigh its disadvantages for most users. Acetonitrile is used in approximately 70 % of all reversed-phase separations.

Tip No.
07

The pH of the mobile phase to too high/too low – what can I do?

The Case

Let us assume that you found the best separation for your particular sample using a C_{18} column at pH 9.5. That might be terrific for you, but your column (if it is not a modern one), most certainly will not enjoy the high pH environment for very long (silica gel dissolves at above pH 8). The column performance will rapidly decrease. Certainly, you could search for a more suitable column such as a C_{18} column with an alkaline base silica gel (see Tip No. 3) or a column material based on a polymer or on aluminum or one of the newest generation (SilicaROD, XTerra, Bonus, AQ...). However, the changeover would require time and money. A pre-column could be the easy way out. Sometimes, you will observe band broadening after the installation of a pre-column, especially for early eluting peaks. Somehow different, but similarly annoying problems arise with low-pH mobile phases, e.g. pH 2 and lower. The C_{18} chains are hydrolyzed, the column bleeds, and the performance decreases. What should you do?

The Solution

You install a pre-column or a short, old C_{18} column, which you will probably find somewhere in the laboratory, between the pump and the injector module (see Figure 7-1).

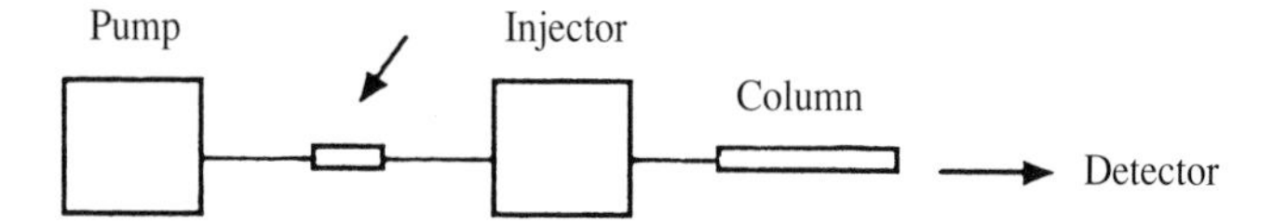

Figure 7-1. Installation of a conditioning column.

What happens? The alkaline eluent is saturated with silica gel and will not disturb the separation column. At low pH, the C_{18} chains of the short column are hydrolyzed, not those of the main column. The installation will also have an additional useful side effect: debris from the pump and other junk will be held back, so that all other HPLC modules will be protected. The choice of the stationary material does not matter – you could even use 100 µm material, since no separation will take place.

Conclusion

This so-called conditioning or saturation column should be in all HPLC equipment – if space is available. It brings only advantages. Since it can be very short, there will be almost no additional pressure.

Tip No.

08

What is the right ionic strength of the buffer?

The Case

Buffered eluents are important for the separation of ionic solutes because they maintain a constant desired pH. Very often, only a defined and constant pH (therefore a buffer) will allow the desired selectivity and reproducibility. So far, so good. Sometimes, you adopt a method with 5 mM salt in the mobile phase, while the next method uses 100 mM or even 150 mM solutions. Sometimes you simply do not know, as in the following example: "Weigh out *x* mg of salt X and add *y* ml of acid Y or base Z!" How critical is the ionic strength and how does the ionic strength influence the separation and column lifetime?

The Solution

Let us first summarize the effects of the adding of salt to the mobile phase:

A. The eluent is more polar, which has two consequences:
   1. Polar solutes elute earlier, they are happier in the more polar eluent and spent more time in it. Non-ionic, lipophilic compounds elute later. Look out for a possible change in the elution order after changing the ionic strength! (see Chapter 5).
   2. Solubility of the polar matrix silica gel increases – "similar eluents dissolves similar solutes". The lifetime of the column decreases either slightly or very much, depending on the stationary-phase material. This is a definite disadvantage.

B. Small batch-to-batch variations of commercial stationary phases are frequently observed. With buffers, the dissociation of available free silanol groups is depressed, thus equalizing stationary phases. The robustness of the chromatographic system increases. This is an advantage.

If you have unknown solutes in your sample, for example impurities or metabolites, which are more ionic than your main sample component, the following can happen: the ionic solutes can literally be swallowed by your alkaline, acidic or metal ions containing stationary phase (see Tip No. 30 and 32). These solutes may disappear totally or elute so late in the chromatogram that they are lost in the noise. If you have a buffer in your mobile phase, this danger is less.

But let us get back to our original question: what does ionic strength do to the separation?

The buffer must be stronger as the danger of pH change increases, or in either of the following circumstances:

- alkaline or acidic silica gel
- injection of strongly acidic or strongly basic solutes

If you separate only weakly polar solutes, a 5–10 mM buffer is often sufficient. Is your method in this case robust? Please check. In all other cases, we recommend an ionic strength of 20–30 mM, especially if you use materials of the 1970s. In ion pair chromatography, an ionic strength of 60–100 mM is usually used. The more homogeneous the surface of the phase (phases of the newest generation), the less important is the strength of the buffer used.

Conclusion

- Using the ionic strength, you can manipulate retention times
- If the pH is important, you have to use the correct buffer to keep this pH constant in order to yield a robust system (see Tip No. 27)
- For strongly ionic solutes, the buffer in an average reversed-phase system should be 20 mM, in a reversed-phase system with ion pair reagents 60 mM or more.

Tip No. 09

How to make sense of the dead volume of an isocratic apparatus?

The Case

You will have certainly have heard about the dead volume of HPLC equipment. If you are one of the users who have measured the dead volume and are on top of it, you do not have to read any further. Simply skip this advice and move on to the next. This section is for those who might think that only theoreticians and academics are interested in dead volumes. They only want a good separation and have no time to play games.

I totally agree with them. What counts in the end is the separation. However, the dead volume does influence the separation, and for early eluting peaks this influence can be considerable. It is therefore useful to look into it more closely.

The dead volume of an isocratic instrument is the volume between the injection point and the detector, excluding the column. It is the volume of connecting capillaries, fittings and detection cell. Since the sample is present in this part of the instrumentation, its total volume should be minimized. You will then have only a small amount of dilution of your sample, thus minimizing band broadening. How do you measure the dead volume?

The Solution

Measurement of the dead volume is very easy and takes only a few minutes.

You have to remove the column and connect the capillary from the injector directly to the detector using a dead volume-free connector. You now have to inject a UV-detectable solute, for example 1 µl of acetone or simply your standard mixture at a low flow setting, for example 0.2 ml/min. If you use a recorder or an integrator, you should set the paper speed at its maximum value for an exact measurement. The injected solute will result in a peak. Please measure the elution time at the apex (see Fig. 9-1).

The dead volume is given by the equation $V = F$ [ml/min] $\cdot$ t [min].

Example: The flow rate setting is 0.2 ml/min and the apex of the peak is at t = 0.2 min. With these values in the above equation, the dead volume can be calculated to be ca. 40 µl.

The question now is: is this any good? Well, within reason.

First rule of thumb: (for the more stringent users among us):

The dead volume of a very good isocratic instrument should be about 1/6 of the peak volume. Assuming a column length of 200–250 mm and a column diameter of 4–4.6 mm, the peak volume is in the order of 100–120 µl. In this case, a dead volume of 20–30 µl is OK.

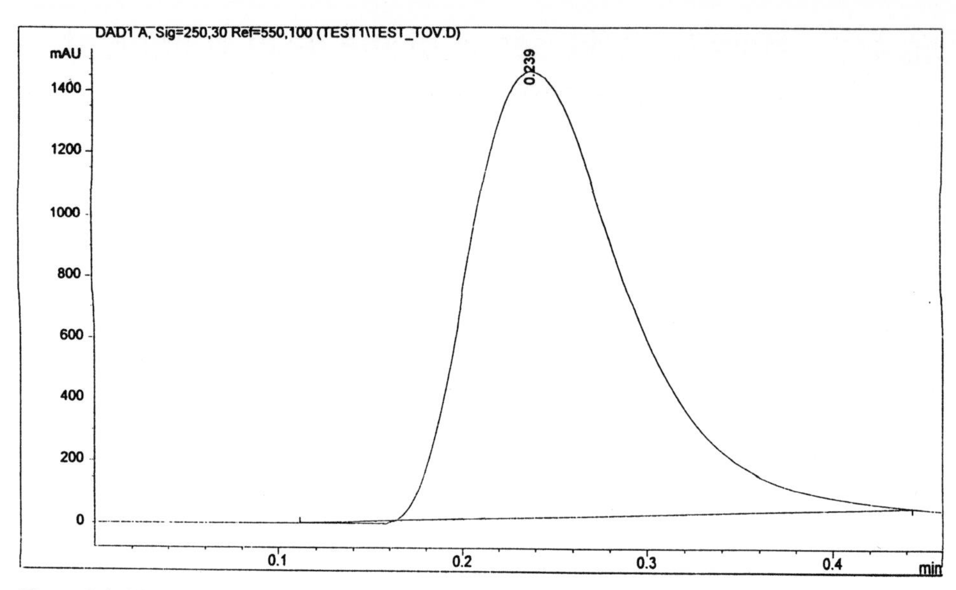

Figure 9-1. Measurement of the dead volume.

Second rule of thumb (for the more generous users):

The dead volume should not exceed the detection cell volume by more than 5–6 times. Using the most common detection cells of 8–12 µl, dead volumes should not exceed 50–60 µl.

Of course, the dead volume is more critical the earlier the peaks elute and the more of them there are. For later eluting peaks and in simple separations, dead volumes are not as important. An example may demonstrate this: Assume we get a lot of peaks separating at 1 ml/min. The first has a peak width of 0.5 min and the last a peak width of 2 min. The peak volume (the volume in which the compound is dissolved) for the two cases is, according the formula

$$V = F \text{ [ml/min]} \cdot w \text{ [min]},$$

$V = 1000$ µl/min · 0.5 min = 500 µl for the first peak and $V = 1000$ µl/min · 2 min = 2000 µl for the last.

Let us furthermore assume the dead volume of the equipment is approximately 100 µl. That means, that of this 500 µl respectively 2000 µl peak volume 100 µl belongs to the dead volume of the apparatus! The additional volume is approximately 20 % in the first case and approximately 5 % in the second.

In addition, dead volumes have more influence if narrow (2–3 mm) and/or short (5–10 cm) columns containing 3–5 µm stationary phase are used than if they are more conventional columns with greater volumes.

An unacceptable dead volume is essentially caused by large-diameter connecting capillaries and most of all by defective fittings. Of course, you can calculate the volume of capillaries, but nobody seems to have the time to do this. The most common ones are therefore listed in Table 9-1 (Source: J. W. Dolan and L. R. Snyder, Troubleshooting LC Systems, Human Press, 1989).

Table 9-1. Volumes of capillary units as a function of their internal diameter (in mm and inches).

Internal diameter		Volume
inch	mm	µl/cm
0.005	0.13	0.13
0.007	0.18	0.25
0.010	0.25	0.51
0.020	0.50	2.03
0.030	0.75	4.56
0.040	1.00	8.11
0.046	1.20	10.72

The left hand column gives possible capillary internal diameters in inches, and the second column gives the conversions to mm. The volumes of the capillaries in µl/cm are given in the right hand column. For example, the volume of a 40 cm long 0.25 mm diameter capillary is 20.4 µl. If you use the common 0.13 mm capillary you are safe, at least concerning the capillary volumes. Be aware of additional capillaries such as a heat exchanger capillary in your detector, that has a volume that you cannot define.

Conclusion

Please measure the dead volume of your isocratic equipment. Before suspecting the column as the cause of a bad resolution, you should rule out the dead volume. The effect of a dead volume that is too large is an unnecessarily low efficiency and broad and/or tailing peaks. My advice on the subject is to use a generally accepted maximum value for the dead volume for all isocratic HPLC equipment in the laboratory. Such a general rule will help in obtaining comparable results from the different equipment and you will obtain similar chromatograms in method transfers.

Tip No.

10

Producing a gradient chromatogram – influence of instrumentation

The Case

A classical case. You produce a gradient chromatogram, and, although you take extra care to stick to all chromatographic conditions used when the experiment was performed with different equipment, you obtain a different chromatogram with your equipment. Why?

The Solution

Almost always, the reason is in a different delay volume of your equipment compared to the equipment the method was developed with. The delay volume is the volume from the mixing point of the two eluents (mixing valve, mixing chamber, T-piece) up to the column. The delay volume can range from a few hundred microliters up to 10 ml depending on the manufacturer. High pressure gradient formation results in general in a lower delay volume than low pressure gradient formation, because the mixing occurs behind the two pumps. Using a low pressure mixing system, the mixture has to pass through the pump first, resulting in a greater delay volume.

While the dead volume of isocratic equipment (see Tip No. 9) results in peak broadening, the delay volume of gradient equipment is the reason why the desired mobile phase composition only reaches the column after x minutes. For example, equipment with a delay volume of 0.5 ml will add an additional time of 0.5 min to reach a defined mobile phase composition using a 1 ml/min flow rate (see Fig. 10-1). This corresponds to an initial isocratic step. Depending on the delay volume, a different mobile phase composition will be present at a defined time after injection, resulting in a different chromatogram with shifting retention time. This situation is schematically illustrated in Fig. 10-1.

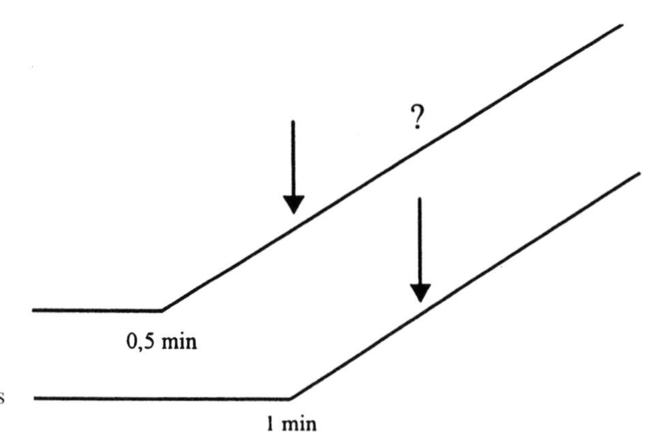

Figure 10-1. Different delay volumes in two types of gradient equipment.

How can I measure the delay volume?
You have to connect two eluents to your gradient equipment:
Eluent A: methanol
Eluent B: methanol + 1 % acetone, to detect the gradient profile with a UV detector

As shown in Fig. 10-2, you run a gradient from A to B. The time lapse up to the detection of the UV-active compound correlates to the delay.

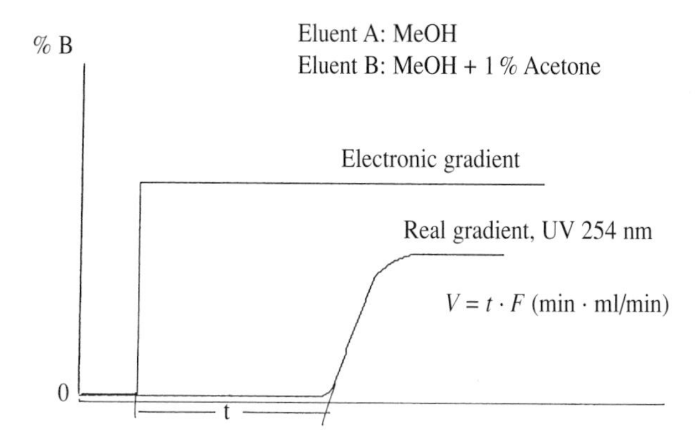

Figure 10-2. Detection of the delay volume.

Because of the delay volume, you have an isocratic step at the beginning of your run before the gradient comes into effect. The delay volume can be calculated according to: Delay volume = time [min] · flow rate [ml/min]; for example: $V = 2$ min · 1 ml/min = 2 ml.

Conclusion

A good SOP for a gradient elution method includes the delay volume of the equipment used! The user adopting the method will be grateful to the author.

As with isocratic equipment (see Tip No. 9), you should agree to a defined delay volume within the laboratory. This will simplify method transfer.

There are two further possibilities to solve this problem:

1. If your software has the capability, you could do a time-delayed injection. For example, if you have a delay volume of 2 ml and use a flow rate of 1 ml/min, you then start the gradient at time point x and inject 2 min later.
2. Start your data collection 2 min later, this should work with most software packages available.

Tip No. **11**

Does the pump work correctly, precisely or accurately?

The Case

This subject is not about hair-splitting; nor is it a joke. Only the true knowledge of accuracy, correctness and especially precision of a pump will tell you the whole story and add to the weight of a quantitative result. What do precision and accuracy mean in this context and how do you measure these parameters?

The Solution

The parameters are measured with a test.
The pump test consists of two parts:

1. Pump accuracy
   Using a stop watch and graduated flask, the accuracy of the volume delivered can be checked (measurement preferably after the column). The question is whether the pump delivers, for example, the desired 1 ml/min or maybe only 0.95 ml/min.
   A bias of 0.5 % is acceptable.
2. Pump precision or pump constancy
   We differentiate between long-term and short-term constancy
   (a) Long-term constancy
   One way to access this is the precision (often incorrectly: reproducibility) of retention times. What is the variation coefficient V_c of retention times using 6 injections of a solute. In other words, how does the flow rate scatter over a longer time interval? With low precision of the retention times, there is the danger of a wrong peak assignment – especially with many small peaks.
   (b) Short-term constancy
   Short-term constancy is a way to describe the flow consistencey during peak elution. It is determined using the variation coefficient V_C of the peak area of 6 injections of a component.

Details of short-term constancy

The area is proportional to the injected amount. If the flow rate actually changes, the residence time in the detection cell changes: at low flow rates, the residence time is longer, at high flow rates shorter. Consequently, the solute absorbs UV light for a shorter time in the latter case, and the peak area is smaller. However, since the injected amount is constant, the product of flow rate and area must be constant:

$m = F \cdot A$; if F decreases, A must increase!

If the area changes, we have proof that the pump does not deliver a constant flow. If you do your data processing using the area, a V_C of 1 % means that you have an error of at least 1 % in your results.

The presumptions for the above made statements are:

- stable solutions,
- reversible interactions,
- constant injection volumes, e.g. with a built-in loop,
- constant pH,
- no measurement on peak slope.

If the flow rate is accurate and precise, we can speak about a correct flow rate. Correctness is therefore the generic term for accuracy and precision. However, in the literature, very often the term "accuracy" is used as an equivalent to "correctness".

One more piece of advice:

In HPLC, concentration-dependent detectors are very common. If the flow rate variations are small, the concentration is virtually constant, thereby having no influence on the signal height. Of course this changes, similarly to the area, when the injected amount changes. This fact allows the following quick decisions:

If the peak height is constant, the injection system is OK

if the peak height *and* peak area are constant, the injection system *and* pump are OK.

Again: long-term constancy does not automatically mean good short-term constancy. See the results of two experiments in Fig. 11-1: Fig. 11-1a shows both good long time *and* short time constancy. Fig. 11-1b shows a measurement with good long-term but poor short-term constancy.

Conclusion

If you prefer to evaluate your data using peak area, a measurement of short-term constancy of the pump is more important than checking the accuracy of the flow rate.

If short-term constancy not guaranteed, a data evaluation using peak height should be preferred over peak area.

Injection	Retention time [min]	Injection	Peak area
1	4.643	1	238.59
2	4.647	2	239.31
3	4.644	3	239.39
4	4.647	4	238.58
5	4.644	5	238.65
6	4.633	6	238.50
7	4.638	7	235.92
8	4.627	8	237.83
9		9	
10		10	
Mean: 4.640	Rel. deviation (V_c) 0.11156	Mean: 238.35	Rel. deviation (V_c) 0.16836

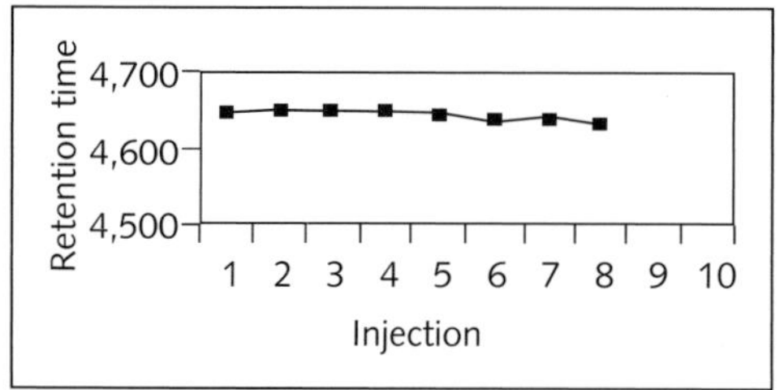

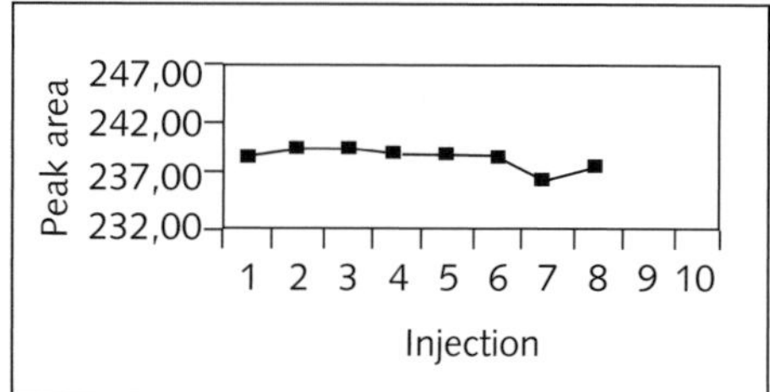

Figure 11-1a. Pump with good short-term *and* long-term constancy; V_c is in both cases excellent.

Injection	Retention time [min]	Injection	Peak area
1	4.643	1	254.33
2	4.647	2	247.23
3	4.644	3	239.39
4	4.647	4	238.24
5	4.644	5	247.39
6	4.633	6	238.55
7	4.638	7	237.48
8	4.627	8	252.68
9		9	
10		10	
Mean: 4.640	Rel. deviation (V_c) 0.11156	Mean: 244.41	Rel. deviation (V_c) 2,66764

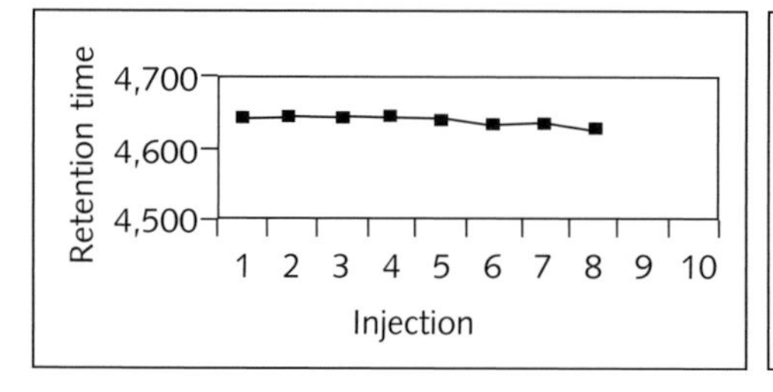

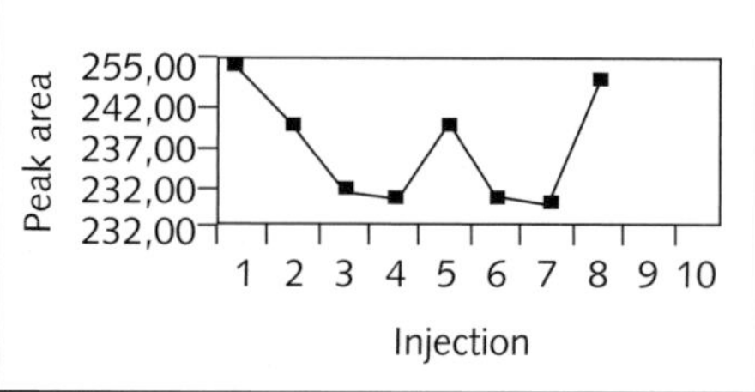

Figure 11-1b. Pump with good long-term but poor short-term constancy (look at the two V_c values).

Tip No. 12

How to test an HPLC instrument and its modules?

The Case

An essential prerequisite for an accurate and precise analysis is well-functioning equipment.

The simplest and safest method for an equipment check is a double injection with a known, stable sample, (control sample, System Suitability Test SST). If the resulting chromatograms are identical with a reference chromatogram which you obtained with the equipment in perfect condition – maybe after a scheduled master check – there is no need for further checks. You know that everything works well (method plus equipment). As criteria for the comparison, you could use retention time, peak area, peak height and asymmetry. Acceptable deviations are to be documented.

If you have any deviation, you will have to check the status of the HPLC modules. Which tests are suitable?

The Solution

Table 12-1 gives an overview of some tests. To simplify the overview, only the most important tests are listed. Checking noise drift and linearity should be part of a detector test, but are not the steps one would necessarily take at the beginning of a search for a malfunction.

Table 12-1. Tests for the various HPLC modules.

HPLC module	Measurement point	Measurement	Recommended limits/ comments
Whole HPLC equipment	System precision	Precision of peak area of a single component using 6 injections out of a sample[3)] solution	$V_{C\ isocratic} \leq 1$ % $V_{C\ gradient} = 2–4$ % V_C is strongly dependent on gradient profile, wavelength, flow rate, etc.
Pump	Accuracy (see Tip No. 11)	Volume delivery	< 5 %
	Long-term constancy	Precision of retention times	$V_C = 0.2–0.5$ % (column oven used, $k = 1–10$)
	Short-term constancy	Precision of normalized peak areas in %	$V_{C\ isocratic} \leq 0.5$ % $V_{C\ gradient}$ 1–3 %, modern gradient systems reach $V_C < 1$ %
Gradient	Gradient precision Mixing quality	Run a step gradient 6–8 times, check gradient profiles	Gradient profiles should not be visible as single traces No “waves”, ”mountains” or other baseline disturbances

3) Some workers use a standard solution. If so, we are talking about an equipment test. If you inject your real sample you make a System Suitability Test (SST). Remember, “system” means equipment plus method.

HPLC module	Measurement point	Measurement	Recommended limits/ comments
Injection system (a) constant injection volume, given by loop volume	Precision	Is given by a complete loop volume	
3 times injection of the same volume	Accuracy	Gives an average, rare mistake in loop	
(b) Variable injection volume	Precision of injection	• To calculate indirect through system precision and normalized peak areas • 6 times injection of a solute and check of peak height	$V_C < 5$ % $V_C \leq 1$ %
	Accuracy	• Double, triple etc. injection volume results in double triple etc. peak area? Prerequisite: pump and detection system OK • gravimetric: e.g. 500 µl of a solute with a known and large specific weight (e.g. bromine benzene), 20 µl injection, one more weighting	0.2–0.5 %
	Memory effect	Injection of pure mobile phase after prior injection of highly concentrated sample (see also Tip No. 33)	If peak-to-noise is smaller than 2:1, there is no noticeable memory effect
Detector	Lamp energy	• Measure the actual response factor at the detection limit • Read energy values with a diagnostic function and compare with the value of a newly installed lamp	
	Accuracy of wavelength	Using solutes with sharp maximum of the extinction coefficient, e.g. terbium perchlorate 218.5 nm, erbium perchlorate 254.6 nm	

Conclusion

The best check in HPLC is always the control chromatogram. Only if you find deviations for previously defined standards will you have to run further tests. All other requirements are out of exaggerated requirements for documentation from out-of-touch QM inspectors or auditors.

Whether it is worth arguing with these people is anyones guess. From an analytical point of view, is it better to check the total system as often as possible. Control charts here are a very powerful tool. With only a little effort, this approach is more valuable than a yearly, but thoroughly technical check of all parts of the system by the service engineer. That is however not to say that the service checks are not important – but trust your eyes and your analytical common sense (if you are allowed!).

Tip No. 13

Injection of solutes as aqueous solutions

The Case

Assume you work with water as the solvent for your sample. If you obtain different peak areas although you inject a constant amount of sample, the reason for this could be a partly irreversible adsorption of your solutes on a "rough" surface in the HPLC system ("hungry surfaces").

The Solution

The irreversible adsorption especially of higher-molecular-weight compounds such as polycarbohydrates and proteins on stainless steel or glass surfaces or on seals is a well-documented problem. If the sample is dissolved in pure water, those surfaces can behave as weak stationary phases (nonpolar pseudo-phases), that adsorb a small amount of the sample. The surface can adsorb sample up to its capacity limit. What can happen now? You inject your standard solution and it is partly irreversibly adsorbed. You will find a peak area corresponding to a smaller amount than is actually injected. The capacity limit of this "stationary phase" is rapidly reached and the peak areas of subsequent injections are correctly measured. However, a comparison with the wrong area of the standard will lead to a wrong quantitative result. We are dealing here with a typical case of a systematic proportional error which prevents you from obtaining an accurate result. What can you do about it?

1. Passivate the equipment with 6 N HNO_3; adsorption on plain stainless steel surfaces is prevented.
2. Using new and very clean equipment, you should inject at least 4–5 times a standard sample mixture to saturate the "hungry" surfaces; use for protein SBA (serum bovine albumin).
3. Approximately 5–10 % organic phase in the solvent for the sample (e.g. acetonitrile or isopropanol, or even dodecylsulfonic acid) are normally sufficient to guarantee that the solutes remain in solution and start interacting only with the "right" stationary phase in the column.
4. If the problem persists, replace your normal injection valve (very often this is the culprit!) with a PEEK valve and, if necessary, also the stainless steel capillaries with PEEK capillaries. Also replace the material of the seals. Lastly you should silanize the glassware you work with.

Conclusion

Note:

Maybe you are not aware that you are suffering sample loss in the injection system. A simple check will tell: inject with a 20 µl loop 40, 60, 80 and 100 µl sample. Your loop has been filled twice, three, four and five times. If everything is OK, the peak area is constant; if the peak increases, a part of your sample is sticking to the injection system.

If at all possible, you should not inject higher-molecular-weight compounds from 100 % aqueous solutions. If this is not possible, perhaps because you have to observe some regulations, you must saturate critical surfaces prior to the injection or try the above-mentioned pieces of advice.

Tip No.
14

What is the largest tolerable injection volume?

The Case

The way you do the sample preparation will result in a strongly diluted sample solution, otherwise you will have to contend with the limit of detection. In any case you would like to directly inject a relatively large sample volume. How large an injection volume can you use without being penalized by noticeable band broadening? We assume that the column is not overloaded; only a possible peak broadening caused by the injection will be discussed here.

The Solution

You can calculate the maximal tolerable injection volume from the variance contributions of all effects to band broadening. This is possible because HPLC consists of stochastic independent processes. However, who has got the time and leisure to do this? Therefore, please find in the following paragraph some numbers and recommendations. Without being given additional explanations just believe it!

Case 1: The sample is dissolved in the mobile phase.
In order to limit band broadening through injection to less then 1 %, the injection volume should only be up 1/6 of the peak volume.
Example:
Using a flow of 1 ml/min, your peak is 0.6 min broad. The peak volume is 600 µl; the acceptable injection volume is therefore 100 µl. If you injected 200 µl, the peak would broaden by 2.5 %. on an older column of 2500 theoretical plates. If you work with a new column with 10 000 theoretical plates, the peak broadening is 9 % compared to a 10 µl injection.
Another rule of thumb:
Do not inject more than 10 % of the column volume.
Example:
You are using a 125 mm × 4 mm column with a flow of 1 ml/min. Column volume can be estimated by the formula: $V = t_m \cdot F / 0.8$ where t_m = dead time and F = flow rate.
To obtain t_m for 4 mm columns use the empirical formula:
$t_m \approx 0.08 \cdot L$ [cm] / F [ml/min]
where $L \approx$ length of the column and F = flow rate.
Here we have $t_m \approx 0.08 \cdot 12.5 / 1 \approx 1$ min.
Going again to the above-mentioned formula we get:
$V = t_m \cdot F/0.8 \approx 1 \cdot 1/0.8 \approx 1.25$
The recommended injection volume should be about 125 µl.

Case 2: The sample is dissolved in a stronger eluent than the mobile phase (e.g. mobile phase: 50/50 methanol/water, sample solvent: methanol) Depending on the mobile phase, temperature, etc. the chromatographic system can take an injection volume of approximately 10–15 µl. However, you have to take fronting and/or a decrease in the retention time into account.

Case 3: The sample is dissolved in a weaker eluent than the mobile phase In this case, there are hardly any limitations. The sample is concentrated at the column head and you will always obtain nice sharp peaks (see also Tip No. 35).

Conclusion

If your sample is dissolved in a weaker eluent than the mobile phase or in the mobile phase itself and you have problems with the detection limit, you can go ahead to inject a lot of solution. Even injection of 200 µl will cause only a 10 % band broadening compared to a 10 µl injection with an almost 20 times increase in peak height. An increase of particular interest in trace analysis.

If using a stronger solvent for the sample than the mobile phase eluent to dissolve the sample, care should be taken and a maximum of 5–10 µl should be injected.

Remember: The more complicated the mechanism of the interaction (ionizable solutes!) the lower the absolute sample should be (app. 0.5 µg).

Tip No. 15

How critical are temperature changes?

Part I: General comments, Detector

The Case

Temperature has a direct influence on the chromatographic result since it influences

- the interaction sample ↔ stationary phase,
- the viscosity of the mobile phase and
- the dissolution of the silica gel.

In addition, several solutes can decompose at higher temperatures. And last but not least, temperature fluctuations in the detection cell have a large influence (electrochemical detector) or simply unpleasant effect (UV detector) on the chromatogram.

The Solution

We can categorically state that a constant temperature increases the robustness in HPLC and is an important aspect for precise results. Therefore, a column oven or air conditioning in a good operating condition is an absolute must for a precise HPLC analysis. Let us take a closer look:

(a) Effects of temperature fluctuations

Reproducibility of retention times	Only at constant temperature are retention times constant and results comparable. As a rule of thumb, a change in temperature by 1°C can result in a 5 % change in *k* values in reversed phase systems.
Quantitative analysis using peak height	A temperature fluctuation also causes a temperature change in the detection cell, thus changing the density of the solution in the cell and in effect the concentration. Since we often work in HPLC with concentration-sensitive detectors (UV, IR, fluorescence), this can have an influence on the peak height: a temperature fluctuation between the analysis of the standard and the last sample in a series can lead to incorrect quantitative results.

Detector — With a UV detector, temperature fluctuation leads to a small drift of the baseline, while for a refractive index detector the drift is large. Depending on the detection method, the effect of constant temperature ranges from very important (fluorescence) to essential (electrochemical detector, conductivity detector).

Effects of a temperature increase

At temperatures of 45–50°C, the danger of bubble formation in the detection cell increases, resulting not only in a higher noise level but also in the appearance of ghost peaks and negative peaks. Finally, at temperatures higher than 50 or 60°C, silica gel is increasingly dissolved and the column lifetime decreases.

Conclusion

In general, a constant temperature leads to increased reproducibility in HPLC. At temperatures above 45–50°C, an increased noise level is observed with an organic mobile phase.

Tip No. **16**

How critical are temperature changes?

Part II: Column, Separation

The Case

Temperature changes lead to changes in chromatographic parameters, both physically (Δ viscosity ⇒ Δ kinetics ⇒ Δ efficiency) and chemically or thermodynamically (Δ adsorption enthalpy ⇒ Δ retention time ⇒ Δ selectivity). Therefore, temperature is a chromatographic parameter which can be used for a rational manipulation of the separation. How can you use the parameter "temperature" in the real world to optimize your separation?

The Solution

An increased temperature has the following effects:

Chemistry

- Retention time and selectivity always decrease (with some chiral separations being rare exceptions). The decrease in retention times is dependent on the solutes and the eluent actually used. Selectivity changes are most prominent for ionic solutes because of the temperature dependence of the pH. This means that, following temperature changes, peaks move at different speeds towards the beginning of the chromatogram.
  But be careful:
  1. It can happen that, at your current temperature, a peak is hiding underneath another peak.
     Temperature changes and repeated injections increase your safety margin on peak purity.
  2. There is a danger of change in the elution order because of differences in the speed of movement.

Physics

- Viscosity decreases result in a pressure drop and a faster diffusion. In addition, faster kinetics, higher theoretical plate numbers, sharper peaks and lower detection limits will arise.

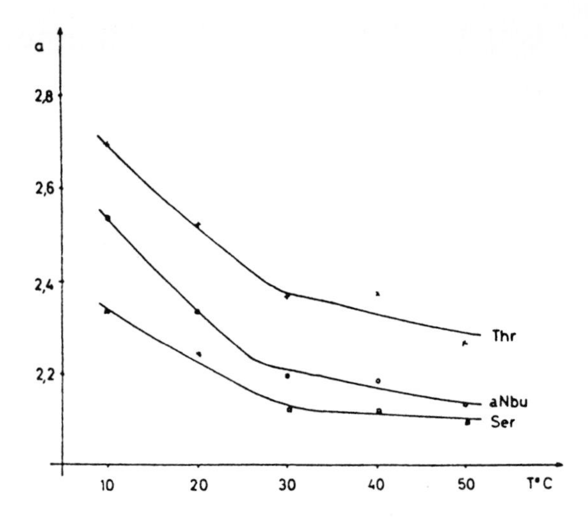

Figure 16-1a. Dependence of the selectivity on the temperature within chiral separations of DL-dansylamino acids.

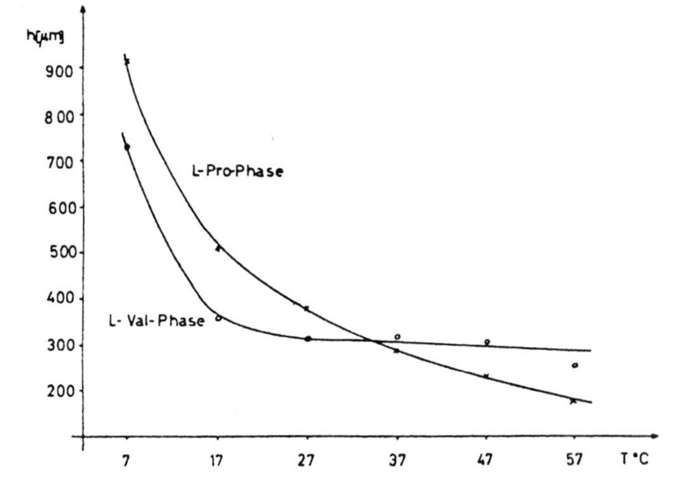

Figure 16-1b. Dependence of the theoretical plate height of DL-dansylamino acids on temperature.

These various influences lead to the following statement:

A temperature increase leads in most cases to a decrease in selectivity and an increase in efficiency. This dilemma is demonstrated in Fig. 16-1, which shows the chiral resolution of DL-dansylamino acids. In Fig. 16-1 (top), the enantioselectivity is plotted against temperature, and in Fig. 16-1 (bottom) theoretical plate height is plotted against temperature. Below 20 °C, the selectivity increases dramatically. Unfortunately, kinetics are rather slow and the theoretical plate height increases rapidly; the theoretical plate number decreases. The decisive factor for the resolution depends on the separation mechanism. With some separation mechanisms with slow kinetics, only high temperatures will result in reasonably shaped peaks: ligand exchange chromatography 40–50 °C, ion exchange chromatography 80–90 °C (see further details below). Figure 16-2 exhibits the enantiomeric separation of racemic barbiturates at 40 °C with ligand exchange using a home-made L-proline stationary phase. If you manipulate the kinetics in your favor, e.g. with a high percentage of acetonitrile in the mobile phase, you should choose a separation temperature of 25 °C because of the better selectivity. This is also shown in Figure 16.3 with the separation of DL-dansylamino acids acids with a 75/25 acetonitrile/acetate buffer, again using the L-proline stationary phase.

Silica gel dissolves more readily at smaller particle size. For 3 μm particles, the effect is noticeable over the approximate range 40–50 °C, and for 5 μm particles over the approximate range 50–60 °C.

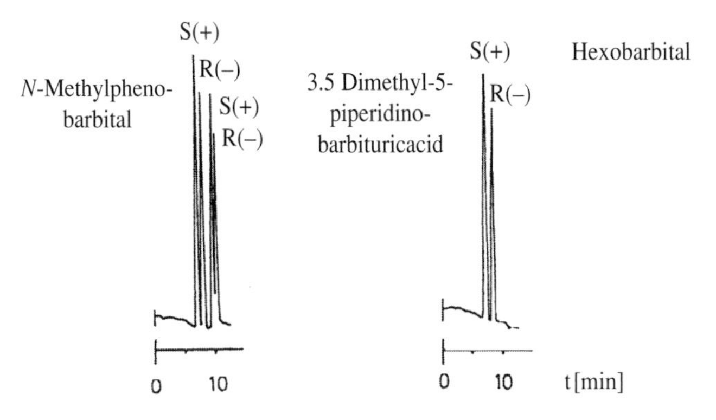

Figure 16-2. Chiral separation of DL-barbiturates on an L-proline stationary phase at 40 °C (eluent: 34/66 w/w acetonitrile/acetate buffer).

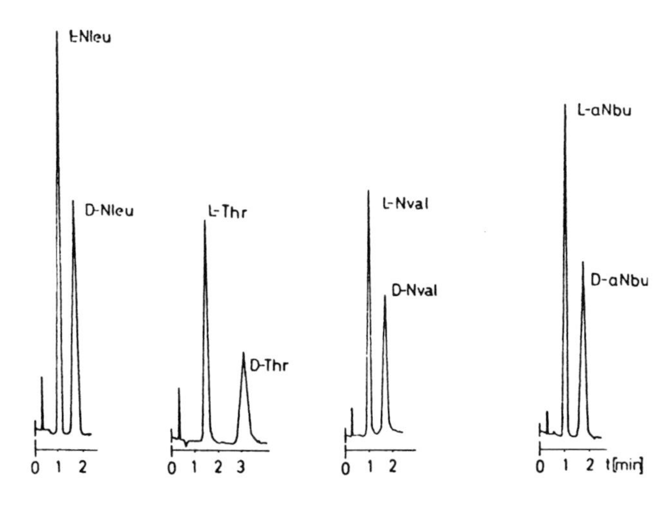

Figure 16-3. Chiral separation of DL-dansylamino acids on an L-proline stationary phase at 25 °C (eluent: 75/25 w/w acetonitrile/acetate buffer).

In summary, a temperature increase will cause:

- an increase of efficiency and mostly resolution, especially for ion exchange
- a decrease in detection limit, selectivity (in most cases), retention time, column lifetime (noticeably starting at 50 °C), pressure and resolution in reversed-phase separations.

Conclusion I

Temperature can be used rationally to optimize your chromatographic analysis:

Temperature increase, advantages:	Analysis speed, sharp peaks, therefore lower limits of detection
Temperature decrease, advantages:	Selectivity improvements (in most cases).

For many reversed-phase separations, a temperature range of 30–40 °C is a reasonable compromise between selectivity and column lifetime on the one hand and efficiency and analysis time on the other.

Temperature is an important optimization parameter whose influence could increase in the near future. This is a direct consequence of the growing importance of certain separation mechanisms. For the interested reader, I would like to discuss this subject in more detail.

In a classical reversed-phase system, selectivity is little affected by temperature. For the separation of similar compounds, which also are of relatively low molecular weight, only small interactions with the reversed-phase surface result in small enthalpy and entropy differences (ΔS and ΔH). This is described by the following two van't Hoff equations, which show the influence of temperature on retention and selectivity.

$$\ln k = -\frac{\Delta H}{R} \cdot \frac{1}{T} + \frac{\Delta S}{R}$$

$$\ln \alpha = -\frac{\Delta(\Delta H)}{R} \cdot \frac{1}{T} + \frac{\Delta(\Delta S)}{R}$$

k	retention factor
ΔH	enthalpy change resulting from the interaction of the solute with the stationary phase. Degree of the strength of the interaction. When the enthalpy increases, peaks elute later in the chromatogram.
R	molar gas constant
T	temperature
ΔS	change in entropy resulting from the interaction of the solute with the stationary phase; describes the degree to which the structure of the solute influences the interaction
α	separation factor
$\Delta(\Delta H)$	difference in enthalpy values for two solutes
$\Delta(\Delta S)$	differences in entropy values for two solutes

For the following separation mechanism, the temperature is essential, because the resulting enthalpy and entropy values are large. These are the separation of large molecules (large $\Delta(\Delta H)$ values) and/or interaction with a structural aspect (large $\Delta(\Delta S)$ values): ion exchange and ion exclusion, gel filtration, affinity chromatography, chelate formation, chiral separations, separations of *cis-trans* isomers.

Some of our own work has shown the following:

1. If the stationary phase and/or the solutes contain cyclic structures, large entropy differences are possible. In this case, the selectivity is strongly dependent on temperature.
2. A steric "relationship" of sample and stationary phase is often more advantageous for selectivity than a chemical relationship. The selectivity is increased if similar structural elements are present in both stationary phase and solutes.

Finally, further advice in this context: If you are interested in an analysis of characteristics of your particular compounds or stationary phases, you should consider van't Hoff plot, which are suitable for these questions. A ln k versus $1/T$ plot results in a straight line with the general formula $y = ax + b$. The slope represents the enthalpy differences and the y intersection the entropy values. We obtained some interesting results using the van't-Hoff plots, but a more detailed discussion is not within the scope of this book. Let us just discuss some of the conclusions that can be reached from this kind of analysis:

(a) $\ln k = f(1/T)$

- Are interactions of polar groups on the separation more significant at a certain temperature?
- Do the "alkyl chains" stand upright at a certain temperature in the case of a highly covered alkyl phase?
- The component is present in two (dissociation) forms
- Both above-mentioned reasons.
- By comparison of the intersection of several straight lines (several solutes), the preference of the stationary phase for certain structures can be determined.

A linear relationship reflects a uniform interaction of the sample with the stationary phase in that temperature range.
A curve or bend means a heterogeneity of the phase as a function of temperature or multiple separation mechanisms, e.g.:

(b) $\ln k = f(\Delta H)$

From $\ln k = f(1/T)$ plots, ΔH values can be obtained for each solute, which can then be plotted against ln k.

- If all values are on a straight line, the interaction type for all investigated solutes with the stationary phase is the same. If the values are not on a straight line, a modified or additional interaction is effective for those solutes.
- If the slope of the straight lines, recorded in different chromatographic systems, is roughly the same, the separation mechanism is the same.

(c) $\Delta(\Delta S) = f(\Delta(\Delta H))$

From this plot, the type of interaction and therefore the driven mechanism for the separation can be investigated, at least qualitatively. This knowledge can be helpful in the choice of the right stationary phase. A straight line parallel to the x axis is an indication of a classical interaction, namely sample $\leftrightarrow$ stationary phase, releasing enthalpy. The larger the slope, the bigger are the entropy differences. In this case, structural differences, e.g. sterically favorable or unfavorable arrangements are important for the selectivity, for example use phases with short polar and long nonpolar groups.

Conclusion II

- If you work with solutes such a steroids, isomers or naphthalene derivatives, or in a separation mode such as ion exchange, affinity chromatography, chiral chromatography etc., you should certainly check the influence of the temperature on the selectivity, but not just in the range 30–40 °C.
- If you would like to have more information about the characteristics of new compounds, van't-Hoff plots can be helpful.

Tip No. 17

How to choose HPLC equipment and a supplier?

The Case

Assuming that in the near future you have to decide on the purchase of new HPLC equipment, which type of equipment will you buy, or how will you help your superiors to make the decision. With the current number of suppliers and types of equipment, this can be a major problem, even for an experienced user. Which questions should you get answers to in the course of the purchase preparation?

The Solution

In this concrete case we assume that the decision that HPLC is the right method for your separation problem was correct – not always an easy decision. You should consider not only scientific-technical requirements, but also general criteria for your purchase.

Among others some general criteria are:

- Sometimes the results of two laboratories are compared such as, e.g., an R + D laboratory versus a quality control laboratory, a laboratory in your country versus a laboratory of a foreign subsidiary, etc. If there is no reasonable case against it, each laboratory should work with identical equipment.
- The experience of the user should fit in with the method and the equipment. An HPLC newcomer should not get started with a method involving on-line sample concentration of a biological sample with double column switching and subsequent post-column derivatization. That is hardly the way to make you happy right from the start.
- How is the service of the supplier in your area? Very often, the answer to this question implies the economic use of your HPLC, because today there is hardly any "bad" equipment on the market, just bad "software" in terms of management, politics of the company etc.
- Choose a specialized supplier. Some companies have their emphasis and their know-how in a certain market segment, such as preparative HPLC, ion chromatography or biochemical analysis.
- Other, more general criteria are budget, sympathy for a company or a sales person, training courses, brand name recognition. For special application, please ask for a customer reference list. And call them!
- Please think about the possibility of purchasing your equipment from a smaller, truly dedicated HPLC company rather than from a large general store for instrumental analysis whose HPLC market is only 10–15 % of his overall business and therefore outside of its main focus.

The technical questions sometimes imply very special demands based on the application. They should also be answered in the preparation of the purchase, for example:

- Can this pump deliver hexane or ether?
- Does the supplier offer a UV detector for my detection requirements at 185 nm?
- What is the linear range of the new fluorescence detector?
- Is the whole equipment inert or only the pump heads and the capillaries?
- Does the pump provides pulsation-free flow to allow the use of an electrochemical detector at 1.2 V?
- Trace analysis: What is the minimum volume of the autosampler vials to provide a required V_C of 3 %?
- How constant is the temperature of the cooling autosampler at 5 °C?

In the end, the anticipated application should decide the overall equipment, accessories and design of the device.

To purchase a compact or a modular equipment is more a question of your personal taste and your priorities. In the following, the pros and cons of compact devices are listed. For modular instruments, the same arguments are true *mutatis mutandis*.

Table 17-1. Criteria for the purchase of a compact HPLC instrument.

Pros	Cons
• Simple and pleasant operation, especially for newcomers in routine analysis	• You are stuck with the quality of the individual modules and cannot choose the best performers available on the market.
• Optimal design, no compatibility problems of the sub-units	• Often increased bench space
• *One* address for technical problems	• Often more expensive than individual modules
	• No flexibility for changing requirements, e.g. is there space and the option to incorporate a second detector and a column switching valve?
	• No flexibility with technical problems, e.g. if only one pump is not working, you cannot switch it with one out of the cupboard and continue working – the whole instrument is out of order

In the following, we would like to discuss some salient points of the equipment in a routine analysis environment and in analytical research (see also "Trends in HPLC" in the supplement section of this book).

But first, one general remark to the provision of your HPLC equipment: if you have a budget of 30 000 to 40 000 US $ or more and your equipment is to be used for more than one application, make sure your instrument is equipped with

- gradient pumping system, possibly capable of micro-bore and short columns operation (high speed, high throughput analysis)
- if possible on-line degassing
- autosampler with injection directly from micro titer plates
- column oven, if necessary for sub-ambient use
- "elusaver" to save mobile phase

The price of an HPLC instrument starting at 40 000 US $ should also include customer training!

Table 17-2. Criteria for the purchase of HPLC instruments for research and routine applications.

Routine analysis	Research
General Requirements: • Simple, robust and service friendly instruments with clear diagnosis functions.	General Requirements: • Purchase a gradient system if you often change your mobile phase; a low-pressure mixing gradient with 3 eluent inlets is advantageous and less expensive than a high-pressure instrument with two or three pumps.
• Micro-bore capability is advised for economic and ecological reasons.	• Micro-bore capability often necessary for analytical reasons.
• A good measurement precision of $V_C = 0.4$ % is most likely more important than a second detector.	• Two detector systems, such as DAD plus fluorescence or MS detector, are often more important than an excellent measurement precision, $V_C = 0.8$ % is often sufficient.
The autosampler should contain a carousel for at least 100 sample vials.	An autosampler is only justifiable if the funds are available. More important is the possibility to inject different injection volumes.
A column oven must be part of the equipment.	A sub-ambient temperature column oven (5–80 °C) improves your options to optimize selectivity.
Pay attention to simple, user-friendly software, e.g.: • Can you run an emergency sample without a major effort? • Change of method or sequence: is this simple or do you need another long learning phase after your vacation? • How many buttons do you have to press to perform an injection ? • Statistics, data storage: is the procedure comprehensible, clear and GLP-conforming (only if necessary)	You should have flexible software with further options regarding the general data processing: with the help of the suitable algorithm, derivative and factor analysis, you can access your peak purity mathematically. The possibility of doing some validation work is also nice to have. Optimization programs are helpful tools – if you make a serious effort to learn them.
	If necessary, use column-switching valves to facilitate selectivity tests with different columns.
Important: Ask at the start of your negotiations about a service contract: how do you rate the response in terms of experience and record?	**Important:** How do you rate the supplier's innovation? Indications are, for example: • applications data bank; high rate of visiting their website • literature with user information (not marketing material!) • how often, where and on which subjects do publications appear?

Are you looking for equipment to satisfy both research and routine application needs? Define your requirements for your needs, buy "research-oriented" equipment and provide for a well-balanced, user-oriented training of your staff.

Conclusion

- Define exactly what you would like to accomplish with your HPLC and who will be the major user. Remember: Only if you know the target can you find the way.
- Define the environment, e.g. how often and from whom you receive samples? Is HPLC the main focus or will it be handled as an add-on?

- Non-scientific aspects could be decisive: service, sympathy/antipathy, track record of the supplier, compatibility with existing instruments etc.
  Remember: "Soft facts" are more important in the long term than "hard facts" – not only in HPLC ...
- Use independent consulting companies for your decision, especially if you are dealing with special fields, such as biochromatography, clinical chemistry, LC-MS, or micro-LC.

Tip No. 18

Is the current method a robust one?

The Case

Your colleagues from the development department send a method to you or your supervisor comes with a triumphant look into your laboratory with a copy of a method in his hand saying: “Do this”. Or you receive the new literature from the company “FantasyChrome Ltd.” with a picture of a chromatogram including the method that is just perfect for you. Before starting something chromatographic, I would advise you to take a step back and check the method in detail over a cup of coffee.

The Solution

With some methods, you know already by looking at them that they can hardly be robust. If you have to adopt a method notwithstanding, you should at least know why it will not work. The method and not you will be the problem in the future. In the following, several typical examples of where you can reckon on problems are listed. This table can of course be extended.

Table 18-1. Some critical points to consider before adopting a new method.

Case	Example	Commentary
Inaccurate formulation in the method:		
	• “sufficiently degassed”	How long? How? (see Tip No. 5)
	• “suitable C_{18} column”	Which? (see Tip No. 2)
	“at room temperature”	In London or Mexico? (see Tips Nos. 15 + 16
	• “the 20 µl loop is over-filled”	Jane uses twice overfilling, John five times. Depending on the viscosity and the concentration of the solution, differences are possible.
Missing parts in the method:	• “with the new DAD XY 2000, we found an error of 0.01 mg for the determination of NOVIAlol in real urine samples”	The absolute error does not tell you anything if you do not know the starting sample amount. A relative error is more informative.
	• “150 × 4.6 mm Supersil C_{18} with pre-column”	How long is the pre-column? Is it directly coupled to the main column or via an intermediate piece?
	• “a 50 % mixture methanol/water”	Weight or volume %?

Case	Example	Commentary
	• "column oven at 50°C"	Which type of oven was used? Air circulating oven or aluminum block – the heat conductivities are different! Also, differences can occur using stainless steel or PEEK capillaries.
	• "the instrument is composed of a gradient pump with an autosampler ..."	What kind of gradient? High pressure or low pressure mixing? (see Tip No. 10)
	• "a 0.05 M KH_2PO_4 solution was adjusted to pH 3 with phosphoric acid"	With 0.1 M or 0.5 M phosphoric acid solution? The acid concentration can result in a different ionic strength!
	• "a volume of 10 µl NOVIAlin was injected at 30°C ..."	What is the solvent for the sample? (see Tip No. 13)
Remarks in the methods pointing to problems with its robustness:	• "Desmethylnoviat elutes from the described 250 × 4.6 mm C_{18} column at 3.5 min, the impurities after 5.5 min using a flow of 1 ml/min ..."	The dead volume of this column is roughly 2.5 min, i.e. the *k* value of the main component is app. 0.4. This is too small. You must expect major fluctuations in the resolution (see Tip No. 41)
	• "195 nm and methanol/water as mobile phase"	Expect a noisy baseline and fluctuating detection limits.
	• "pH = 6.7" "Mobile phase A :12 mM KH_2PO_4 "19 ml DMSO in the mobile phase"	There is nothing a priori wrong with uneven numbers. However, they are suspicious. Check the method validation closely.
	• "Mobile phase A: 0.05 M phosphate buffer pH 7.5, mobile phase B: acetonitrile, gradient 10 % ⇒ 90 %"	This is not only a gradient in elution strength but also in terms of pH and ionic strength. Assuming further unfavorable conditions (matrix, temperature). The method robustness could suffer.

Conclusion

The examples above are designed to make you pay attention to chromatographic details, to explore what and why things can go wrong. Of course you can rely on "exotic" or even "forbidden" methods. But this is beside the point. I have had only the best results in terms of selectivity improvement with the methods listed below. I quickly needed a selective method for a good separation. My intention was not the evaluation of a robust routine method. Often you have to decide on your priorities in terms of selectivity, detection limits and robustness.

- With a silica gel column covered with Cu^{2+} and with water as the mobile phase we obtain a good separation of free amino acids.
- With an eluent containing 99 % H_2O, 1 % isopropanol and 0.1 M tetrabutylammonium chloride we obtain good separation of ionic compounds on a C_{18} column without tailing.

- With classical GPC-columns and organic eluents we could separate low-molecular-weight compounds (no exclusion possible !!!) extremely well – I do not understand this, but it works!
- Water as eluent and a diol phase give you good resolution of polar compounds.
- With a 30 % acetic aqueous solution as eluent and a C_8 RP phase we get a good separation of very polar compounds.
- A silica gel column in series after a standard C_{18} column increases the selectivity for polar compounds with similar polarities (MeOH/H_2O eluent).

3. Problems and their Solutions

Tip No. 19 Sample preparation – how critical are which mistakes?

There are those happy users who dissolve their sample in the mobile phase and inject right away. If you are one of these and you also have isocratic conditions with acetonitrile/water, you can forget about sample preparation for now. There are however those users who need to quantify minor unknown impurities of a metabolized metabolite from bile fluid and others who hunt after heterocyclic sulfur-containing compounds in crude oil residue from Iran and compare it with Iraqi oil. They all have my sympathy. In the following table, a few typical errors occurring in a "normal" sample preparation are listed.

Table 19-1. Typical sources of error in sample preparation.

Problems	Solutions/Advice
• Difficult to dissolve samples are acidified and/or heated and directly injected: you should anticipate a precipitation since the stationary phase is a good solid catalyst.	• Use acidic mobile phase and/or run your separation at elevated temperature • Cool down, filter and only then inject • Never inject a saturated solution!
• Sample adsorbs irreversibly on glass container, stainless steel capillaries, filter, etc.	• Use blank samples (placebos) for analysis; determine recovery rate. • If necessary, saturate surface with sample before proceeding.
• Is the sample solvent stronger than the mobile phase (e.g. higher amount of acetonitrile)? Fronting, double peaks and retention time shift are possible.	• Injection volume ≤ 20 µl • If necessary, dilute sample solvent with mobile phase • Co-inject solution mediator (e.g. 2 µl cyclohexane)
• Sample is dissolved in ultrasonic bath. Here, new solutes might arise, sample components might change (formation of radicals, thermal decomposition).	• Optimize temperature and time of ultrasonic treatment. Most problems appear at a frequency of > 50 kHz and ultrasonic treatment of > 20–30 min. • Is microwave an alternative for you?
• Inaccurate method description I: e.g. "... dissolve x and fill up to 100 ml..." What can happen? – Solute is 100 % dissolved, everything OK – Formulation contains precipitated parts Results: constant systematic error – you always find more.	• Check method robustness during the method development stage, for example by checking laboratory precision.. • Before routine use, ask an independent expert to judge the method description for real life use.

Table 19-2. Errors caused by preparation of mobile phases.

Mobile phases	Possible Mistakes	Solutions/advice
(a) Water/acetonitrile or methanol mixtures.	• Dilute *x* ml water to 1 l with methanol or vice versa – volume contraction.	Constant, documented procedure.
	• no or insufficient degassing.	Degas (see Tip No. 5).
(b) Old mobile phase	• THF can have peroxides.	Determination of the stability in your own laboratory conditions (temperature, supplier, storage conditions, etc.).
	• with acetonitrile, traces of propionitrile and methacrylnitrile can lead to "ghost" peaks in gradient runs.	Put a small Al_2O_3 column in the MeCN capillary to give on-line filtering of your MeCN.
(c) Use of buffers	• not filtered. Danger: plugging the system	Filter phosphate buffers with 0.45 µm filter, acetate buffers with 0.2 µm filters.
	• Concentration of microorganism in mobile phase/instrument too high Danger: plugging, increase back pressure, ghost peaks.	Inhibit growth, for example with " 0.01 % sodium azide in mobile phase. Flush with a 3 % solution of H_2O_2.
	• Inaccurate method description II: "... adjust pH to 3.5 ..." After or before addition of methanol? Danger: different pH of mobile phase	Consistent, documented procedure. Always do the same things in the same way.

A few further points to remember

1. If you have the possibility, please check the following or try, if necessary to influence them:
   - Is the procedure for getting the sample documented/validated?
   - Has the stability of samples and mobile phases been checked?
   - Sample transport, storage– is their influence on the samples known?

2. Using sensitive methods such as ion pair chromatography, small changes in sample preparation can lead to negative peaks and ghost peaks.
   Things that you could change:
   - Detergent used for cleaning laboratory glassware
   - Water quality, conductivity, pH
   - Storage, temperature, humidity
   - Supplier of ion pairing reagents

3. You "quickly" centrifuge a sample and put the vial in the autosampler.
   Problem: For the 6th injection, you obtain a smaller area compared to the 1st injection; in addition, the injection needle is often plugged.
   Cause: Precipitation of solutes with time, the absolute amount of sample in solution decreases.
   Solution: Optimize sample dissolution.

Tip No. 20

Flushing of an HPLC equipment

The Case

Flushing is certainly one of the best methods to prevent and to solve problems in HPLC. For most users a flushing step is simply part of a perfect gradient method. In isocratic elution, flushing is used at the end of a measurement series, to change to another method or merely when necessary, and is also advisable for economic reasons. Let's take a closer look at the symptoms that indicate the need to flush, what the underlying causes are and how best to flush.

Note: Some of the symptoms listed below (tailing, change of retention time) can also have other causes..

The Solution

Table 20-1. Flushing of HPLC equipment.

Question	Answer
When is a flushing step necessary?	1 High pressure 2 Tailing, peak deformation, double peaks 3 Noisy baseline 4 "Memory" effects, ghost peaks, negative peaks 5 Change in retention times
What are the causes and their symptoms (see within the parentheses).	– "Dirt" (high-molecular-weight molecules, microorganisms, dust, salt, pump oil (!), undissolved matrix components, silica gel etc.). This great unknown plugs or covers the most vulnerable areas: frits (1 and 2 above), injection needles (4), stationary phase (1, 2, 4, 5), detection cell (3) etc. – Tough solutes (in general high-molecular-weight components) "gluing" to the stationary phase and stainless steel or glass surfaces (1, 2, 4) – Metal ions on the stationary phase often cause tailing by forming complexes (see Tip No. 32)
Now let's get to the flushing.	
What do we want to remove?	*With what should we flush?*
• Inorganic impurities, salts	Water or better ca. 0.1 N HNO_3 (pH not under 2.0 since a lot of endcapped phases are sensitive to hydrolysis, see Tip No. 7)
• Organic impurities	In order of efficiency: methanol or acetonitrile, isopropanol (with the best flushing/efficiency ratio), $CHCl_3$, THF, hexane or heptane. As a last resort: DMSO acidified with 0.1 % HCl

Question		Answer
• Metal ions		10–20 mM EDTA
Algae, (dead) microorganisms		6 N HNO_3 and/or 3 % H_2O_2 solution for the whole instrument, NH_3 conc. or acetone for a local cleaning
Where should it be clean?		*Flushing procedure*
Pump, autosampler, capillaries, detection cell		Hot water, isopropanol, 6 N HNO_3 (followed by water flush to neutral)
Injection valve		With detergent followed by an extensive water flush, isopropanol; algae, fungi and bacteria with NH_3 conc. or 6 N HNO_3.
Column:	– C_{18}	Depending on the problem as described under organic and inorganic impurities, see above, possibly to switch flow direction.
	– SiO_2	THF, CH_2Cl_2, hexane
	– GPC columns (polystyrenedivinylbenzene)	Toluene or THF, 1 % mercapto acetic acid in 100 ml toluene or THF
	– Ion exchanger, chiral columns, speciality columns	Since those stationary phases can have very different characteristics, strictly follow the recommendations of the manufacturer.

Conclusion

Using pure (!) water, an appropriate equilibrium step[1)] and – if necessary – filtration of the sample are adequate measures for gradient elution to avoid "dirt" and ghost peaks. With isocratic elution, act only when necessary (see above), however some precautions such as filtration are right on target (see Tip No. 19).

Remark 1: The necessary flushing volume per step is ca. 10–15 column volumes, which means 20–30 ml for the common 15 cm and 20 cm columns or a flow of 2 ml/min for ca. 10 min.

Remark 2: It can happen that your column is clean for the first time in its life after a radical flushing step. Some impurities of the stationary phase responsible for the selectivity are now gone and the separation changes.

1) Rule of thumb for the equilibration time between gradient runs:
$t_E \approx 0.1 \cdot t_m \cdot \Delta\%B$
$t_m \approx$ dead time
$\Delta\%B$: difference in B from beginning to end of the gradient run.
Example: 125 mm × 4 mm column, flow 1 ml/min, gradient 20 % B to 80 % B
First we have to measure or estimate t_m:
$t_m \approx 0.08 \cdot 12.5/1 \approx 1$ min (see Tip 14)
Then we can calculate t_E according to the above formula
$t_E \approx 0.1 \cdot 1 \cdot 60 \approx 6$ min.

Tip No. 21 Dirt in the UV detection cell

The Case

Depending on the sample and/or the matrix, dirt can cover the surface of a UV detection cell. Consequences are:

- A part of the light energy is lost and the baseline is noisy
- The dirt layer increases the formerly smooth surface area. The UV detection cell becomes an air collector and small air bubbles, possibly present in the mobile phase, are restrained and grow to a larger air bubble, resulting in spikes, noise and baseline drift and jumps.

In summary, dirt in the UV detection cell causes inferior peak-to-noise ratio, spikes and baseline problems.

The cell needs cleaning. How can this be done in the most efficient way?

The Solution

A dirty UV detection cell can be cleaned by direct injection of

- detergent
- hot water
- isopropanol
- 6 N HNO_3 (with a subsequent extensive flush of water)

using a common 10 ml syringe.

If the cleaning step is unsuccessful, an optician's trick might help :

Take a saturated solution of collodion in ether (on sale in pharmacies) and apply it carefully to the dirty lenses, windows, mirrors, etc. After two minutes, the ether is evaporated and the white layer can be simply peeled off. The optics is clean again. This is a thorough cleaning method for all optical parts except the gratings. These cleaning solutions are available as ready-to-use cleaning kits from various suppliers (opti-clean polymer). However, after using this cleaning procedure you lose ca. 10 % of the light energy.

Conclusion

Avoiding mistakes is certainly easier on your nerves than fixing them. If the above-described problems are persistent, you should take a look at your sample preparation and the mobile phase components and solvents used. Unnecessary down time can also be avoided if the UV detection cell is cleaned on a regular basis, which does not take much time.

Incidentally, by using a quality card (control chart) your professional eyes will immediately see if it is time for the next routine check and preventive measure. You could enter the actual peak height as a function of time (day, week), the ratio of peak area to peak height, maybe at two wavelengths. This can show up interesting trends.

In general, there should be more use for SPC (statistical process control) in the laboratory. With the current commercial software programs it is easy to get the data directly from the equipment. SPC is a simple universal useful tool for monitoring data, enabling early recognition of errors and failures and observing trends. The point is that you get much more information, besides just statistical data such as relative standard deviations and mean values.

Tip No. 22

The lamp is new – what happened to the peak?

The Case

You notice that compared to chromatograms recorded earlier, your peaks are smaller or simply appear to be smaller. There could be many causes and we only list some of them:

- injection error
- irreversible adsorption on stationary phase or somewhere else
- a different mobile phase (eg. pH drift)
- unstable sample
- wrong integration parameters
- higher flow
- differences in temperature lead to difference in pH

If you can exclude these errors, the detector is to blame – but what exactly is wrong with the detector?

The Solution

The following are possible causes:

1. Detector settings
Could it be that you changed the detector settings? Wavelength, breadth of slot, sensitivity, time constant?

2. Leakage in the detection cell
If the baseline seems to be OK and the chromatograms appear to be not too noisy but the peaks show tailing, a leak directly from the detection call might be the reason. A hair crack is easier to spot with an acetonitrile/water mixture than with the more viscous methanol/water mixture. Using methanol/water, the mixture passes smoothly over the crack. Larger peaks compared to acetonitrile/water mixtures are the result.

3. The lamp is too old
If your deuterium lamp is old, it can be checked by the reading of the hours of operation. A critical limit is 800–1000 hours – in theory. However, it may well be that you can work very well after more than 1500–1800 hours. An old lamp can be detected by from the noisy baseline and small peaks, which should not be confused with spikes. Spikes are sharp lines due to air bubbles. Modern detectors contain a counter for lamp operation hours and also indicate the actual lamp energy. It should therefore be quite straightforward to identify an old lamp as the problem. Remember the control charts (see Tip No. 21).

Note that if a UV detector is used, peak areas should be independent of the age of the lamp, since only relative values are measured. "Only" the peak-to-noise ratio deteriorates and the limit of detection increases. A decrease in lamp energy can also lead to a decreased linear range of the detector.

4. The lamp is defocused
This will also lead to a decrease in lamp energy. The solution is to focus according to the manufacturer's manual; often you only have to turn some adjusting screws to optimize the lamp energy.

5. Wrong wavelength
You can check whether the displayed wavelength is correct with the aid of filters. Some manufacturers build these into their instrument. You can also use a test compound, such as erbium perchlorate, which exhibits a sharp maximum at a defined wavelength with λ_{max} = 255.5 nm.

6. Old grating with "spots"
No comment is necessary: an old grating should be replaced. This will be necessary less frequently than a lamp replacement, but an experienced service engineer will recognize the problem.

7. Dirt in the detection cell (see Tip No. 21)

8. Dirt on the mirrors
Dirty mirrors contribute to a noisy baseline and an unfavorable signal-to-noise ration. For cleaning see Tip No. 21 or ask your manufacturer.

Conclusion

The most important tools for identifying a detection error as the cause for unsteady peaks are:

- A well-documented service protocol, not only covering the chromatographic conditions but also detector settings
- a control chart
- diagnostics function on the instrument and
- a reference chromatogram with a reference solute (working standard, reference standard, control sample).

Tip No. 23

What are the causes of pressure changes or deviations?

The Case

Pressure changes in the chromatographic system, together with shifting retention times, are the most prominent day-to-day problems. The causes can vary, originating in both the mechanics and the chemistry. Which are the most likely causes?

The Solution

Extensive lists of possible causes and their solutions can be found in the literature, e.g. in Dolan, Snyder "Troubleshooting LC-Systems" (see literature in the Appendix).

Table 23-1 is intended as a quick help in the daily routine. To simplify things, only the most common errors are listed, rare or more theoretical causes are not mentioned. The list reflects a personal choice.

Table 23-1. Pressure changes in HPLC – the most common causes.

Observation	Other indications	Possible causes	Further help: what happens to the dead volume t_m?	Commentary/clues
$P\uparrow$		• Flow increase	$t_m \downarrow$	
		• System plugged	t_m = constant	
		– Column		Often at the column head, reverse flow direction and flush
		– Frits		If necessary clean (isopropanol, acetone), replace
		– Precolumn		Replace
		– Capillaries		Salt precipitation more common with 0.1 mm capillaries
		• Δ Eluent	t_m = constant	Example: switch from acetonitrile/ water to methanol/ water; higher pressure because of higher viscosity (see Tip No. 6)
		• Smaller particle size	t_m = constant with same packing density, otherwise $t_m \uparrow\downarrow$	Change from 7 µm to 3 µm pressure drop increase 4 times
		• Temperature decrease	t_m = constant	Pressure increase because of increased viscosity
		• Longer column	$t_m \uparrow$	
		• Transducer OK?		Pressure reading a zero flow? If at $F = 0$ you have not also $P = 0$, new adjustment (small screw, often at front end of pump) or replace pressure sensor
$P\downarrow$		• Smaller flow rate	$t_m \uparrow$	
		• Totally "wrong" mobile phase	t_m = constant	
		• Leak in system	$t_m \uparrow$	
		• Pump head seal defect	Δt_m	
		• Air	Δt_m, most $t_m \downarrow$	
		• Transducer defect		Pressure reading at $F = 0$? etc. (see above)

Observation	Other indications	Possible causes	Further help: what happens to the dead volume t_m?	Commentary/clues
ΔP	permanent small fluctuations	• Air bubbles		
		– in the pump		(a) Purge/prime (b) Flush with isopropanol; better degas in the future
		– frit of suction element in mobile phase container clogged		Clean suction frit with water, acetone or 6 N HNO_3, replace with suction frit made from PEEK/plastic and 10 µm pores
		• pump head seal defect		
				Pressure sensor defect
	Pressure increases, decreases, increases etc.	• Inlet valve(s) is (are) clogged; one pump head is malfunctioning	Δt_m, also flow fluctuations	
	Working with a gradient	ΔP for gradients to be expected	t_m = constant	Water/methanol gradient: viscosity and pressure maximum at ca. 60 % methanol Water/acetonitrile gradient: constant decrease in viscosity and therefore pressure
$P = 0$	No flow	• Pump stops because of high pressure limit • Pump motor defect • Broken pistol		
		• "Too much" air in the pump		Purge extensively, prime by forcing mobile phase through pump with syringe
	Flow	• Transducer not OK		Pressure sensor at $F = 0$ etc. (see above)

Conclusion

If the pressure changes, a check of the flow rate indirectly with the measurement of t_m is a good way to pinpoint the cause: flow or "chemistry"? Also, the diagnostics functions of modern devices, such as leak sensors should be used.

Tip No.

24 Is the right or the left pump head defective?

The Case

The inlet and outlet valves are the prime weak points of each HPLC pump. If you have noticed flow fluctuations and ruled air out as a possible cause, you are left with the unpleasant task of taking the pump head apart. Many of today's pumps are double headed. The question is which pump head to take apart, the right or the left one? In addition, is the inlet or the outlet valve clogged. A 25 % chance to take the right part apart is not bad, but a simple trick will allow you to improve your chances.

The Solution

- You suck a 1 cm air bubble through the Teflon tube up to 15 cm before the inlet valve of the primary pump head.
- The pump is then switched on and the flow set to 2 ml/min.
- At *normal operation*, the air bubble moves with each pistol movement in the direction of the inlet valve.
- If the air bubble moves rhythmically back and forth, the *inlet valve is defective.*
- If the air bubble does not move at all, the *outlet valve is defective.*

Conclusion

There is another case (see Tip No. 35) where air can come to the rescue instead of causing us problems.

Tip No. **25**

Baseline noise and damping

The Case

Your pump is pulsating extremely badly. Although you are using new valves and seals and a more completely degassed mobile phase and have thoroughly cleaned with isopropanol and replaced the transducer there is really no more you can do! The situation does not improve. You would still like to keep the pump, but you are trying to carry out trace analysis and need a good signal-to-noise ratio and/or you are working with a pulsation-sensitive detector such as an RI detector, a fluorescence or an electrochemical detector. What next?

The Solution

Very simple: you install a good pulse damper behind the pump in your system. That is certainly not new, but I have the feeling that many users simply forget that this relatively low investment perfects any pump.

However:

1. This possibility is only for isocratic separations or sometimes for high-pressure gradients (damper before mixing chamber). Because the dead volume of the damper is several ml, the delay volume for low pressure gradients would probably be too high (see Tip No. 10).
2. You have to use a really good damper such that supplied by Shodex for ca. 1000 US $.

To test the efficiency of the damper we have deliberately tested with a single-head pump. This type of pump is known for its large flow variations. Test by switching off the pump; its noise contribution is easily measurable. This way, we can assess the influence of the damper.

Figure 25-1 shows the baseline at the most sensitive setting of a UV detector. At 4×10^{-5} AUFS, the baseline is very good, much better than that of many double-head pumps, which we tested in parallel.

Figure 25-2 shows the separation of four aromatics with a 2 mm micro-bore column using a flow of 220 µl/min. This separation is quite good.

We wanted to be even more strict and used an RI detector, known for being more flow sensitive than a UV detector. The baseline was a straight line. Subsequently, we zoomed in on the baseline (3.96 to 4.14 min and 5.56 to 5.62 mV) and still did not see any fluctuations.

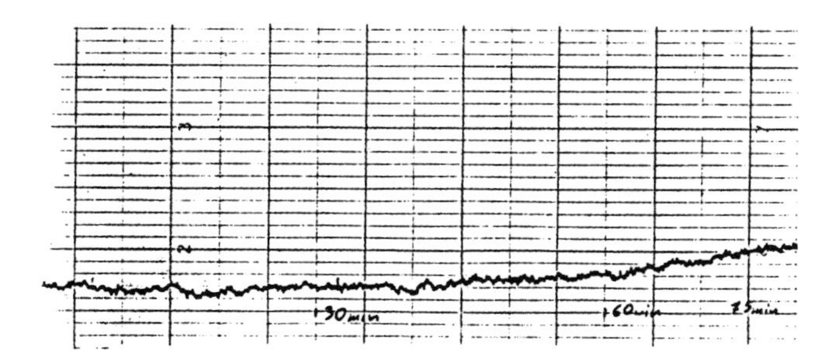

Figure 25-1. Baseline of a single-head pump with an additional damper at 4×10^{-5} AUFS.

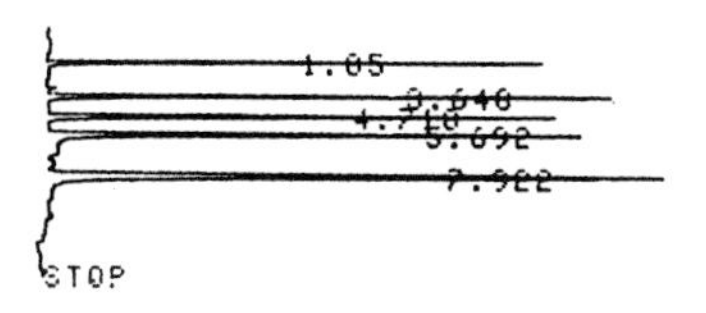

Figure 25-2. Separation of four aromatics with a 2 mm micro-bore column using a flow of 220 µl/min.

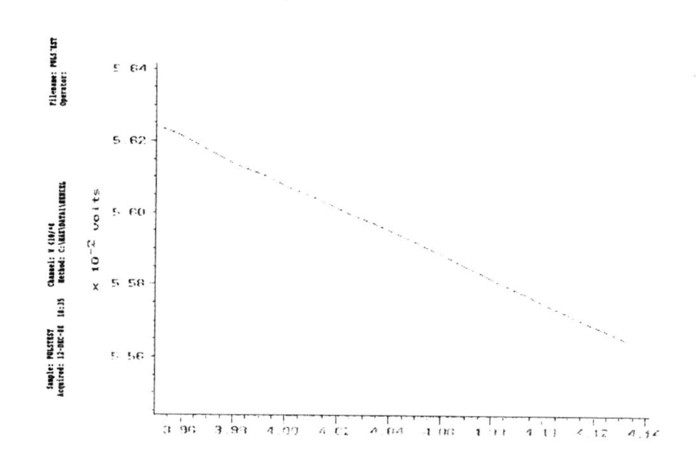

Figure 25-3. Baseline of an RI detector, very much 'zoomed in' (see text).

Conclusion

If for any reason you need a very stable baseline for isocratic separations, remember a damper behind the pump. The result will surprise you. Remember, old solutions are often still good solutions. An additional good possibility is the use of electronic dampers to limit electronic noise due to rear electrical connection or when using sensitive detectors e.g. electrochemical, amperometric or radioactivity detectors.

Tip No. 26

The retention times increase – is it the pump or the mobile phase?

The Case

You are probably familiar with this annoying and all too frequent problem: retention times shift – often backwards. The causes for this can vary (see Tip No. 29):

- air in the pump head
- dirt at inlet or outlet valves
- leakage

} these causes will result in a lower flow

- decrease of column temperature
- changes in the stationary or mobile phase.

In order to pin down the problem quickly, we have to ask whether we can exclude one or more causes mentioned above with respect to both chromatograms (see Fig. 26-1).

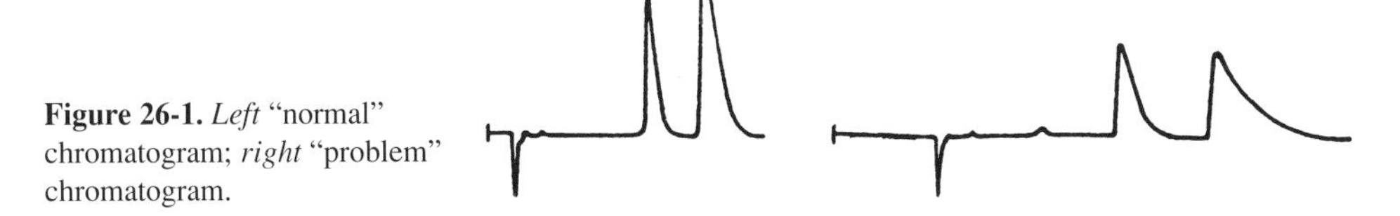

Figure 26-1. *Left* "normal" chromatogram; *right* "problem" chromatogram.

The Solution

You obtain the solution from the chromatogram. A change in temperature, stationary or mobile phase, which includes additives as well as the percentage of the organic part (methanol or acetonitrile), has no measurable influence on t_m. This is only influenced by the flow rate. In the example shown, the cause is a lower flow. Leakage or flow inconsistency of the pump, e.g. a shift of t_m, always results in a change of flow rate – certainly keeping column dimensions and packing density constant.

Conclusion

If you are facing unexpectedlylong retention times, check t_m. The value of this tells where to start searching for the error.

In the search for the cause of retention time changes, we can make several distinctions:

If the dead time t_m is constant and the retention times increases, this means that the pump is OK and the cause is temperature, column or mobile phase.

If the dead time and retention time increase, the cause is a malfunctioning pump or a leakage; in this case you do not have to prepare a new mobile phase!

If you do not find t_m in your chromatogram – which is a pity (see Tip No. 41) – you will make progress by answering the following question:

What happens to the peak area during the retention time increase?

Observation	**Conclusion**
Peak area is constant	Flow is constant
Peak area increases	Flow has decreased since at constant injection amount m (the product of flow F and area A) stays constant (see Tip No. 11) $m = F \cdot A$

Criteria and Causes

Dead volume and area stay constant, retention times decreases ⇒ Cause: ΔT, Δ column or Δ mobile phase.

Dead volume and retention time increases ⇒ pump is defect or you have a leakage, e.g. a decrease of flow.

Tip No. 27

Which buffer is right for which pH?

The Case

It was desired to develop an HPLC separation of neutral and ionic components in one run for routine use. The selectivity was sufficient at first (selectivity factor α = 1.2). However, the user complained increasingly of problems with the reproducibility of the retention times, although the column was sufficiently equilibrated. The chromatographic system used isocratic conditions with a UV detector and a commercial C_{18} column. The mobile phase consisted of 40 % methanol and 60 % 40 mM phosphate buffer (w/w) at a pH of 4.5. What was the problem?

The Solution

As the user had chosen an inadequate buffer, the pH of the mobile phase could change. Why? Buffers are used to keep a predetermined pH constant. Buffering however is only "guaranteed" in a pH range of ± 1 pH unit of the pH of the acid/base pair used. For phosphate buffers, this means ± 1 pH units around its pK_a values of 2.1, 7.2 and 12.3. Only in this range can we be sure that the pH of a phosphate-containing mobile phase and therefore the retention time of the polar solutes remain constant. Weakly polar solutes and a constant pH of the injected sample rarely cause problems.

Conclusion

The buffers recommended for robust mobile phases are listed in Table 27-1 below:

If you would like to work around pH 6, you can choose either an acetate or a phosphate buffer. Because of its better UV characteristics I would choose the phosphate buffer unless it is to be used with LC-MS coupling. (Some modern MS systems can also work with a phosphate buffer.)

If you want to work between pH 3.3 and 3.8 and a constant pH is very important, you can use a citrate buffer. However, if you can avoid this, please do; your pump might be grateful for it. For the weakly acidic region you could also use a formate buffer.

Table 27-1. pH values and suitable buffers.

Range	pH (approx.)	Buffer
Strongly acidic	1.3–3.3	Phosphate buffer (dihydrogen salt)
Weakly acidic	3.8–5.8	Acetate buffer
Neutral and weakly alkaline	6.2–8.2	Phosphate buffer (hydrogen salt)
Alkaline	7.0–9.5	TRIS TRIS: hydroxy methyl amino methane) or borate buffer

Tip No. 28 An interesting alternative for the separation of acids and bases with a buffer (for a detailed discussion on the separation of acids and bases see Chapter 5)

The Case

At an acidic pH, weak acids are not dissociated. In their undissociated form, an interaction with a C_{18} phase is possible and the retention time is large. However, bases are protonated in their ionic form, an interaction with the hydrophobic surface is unlikely, and they elute early in the chromatogram. The situation is reversed in the alkaline range: bases are not dissociated and acids are ionized. This is the reason why the selectivity of acids and bases can be manipulated by the pH of the mobile phase.

Does this also work without a buffer?

The Solution

Figure 28-1 shows the dependence of retention time on methanol content in the mobile phase for the separation of organic acids and bases with an reversed phase. The plots are very interesting.

From 100 % to ca. 70 % methanol, retention times of all sample components are small. From ca. 50 % water, the ionic character of the solutes contributes more strongly. We obtain similar results without buffer and with an alkaline buffer. The strong base *m*-toluidine is dissociated and strongly retained by the stationary phase. Next, the weaker, undissociated base aniline elutes. The weak acid, benzoic acid, is very little dissociated and an interaction with the stationary phase is possible. The stronger acid, formic acid, is present only in its ionic form, so that an interaction with the reversed phase surface is not possible. The ionic character also shows up in the peak form of benzoic acid in comparison to aniline (see Figs. 28-2 and 28-3). This curious peak form is also typical of other acids such as phthalic acid.

Conclusion

In 20–50 % water/methanol mixtures, a separation of weak acids and bases is possible without a buffer addition. In order to obtain a non-tailing peak for the analysis of basic solutes, make sure you use the suitable C_{18} column (see Tips Nos. 43 and 44).

Use this trick if you really must work without buffers or if you want to compare stationary phases with each other; without a buffer the differences are more visible. If not, use buffers. The reproducibility of the separation will be better.

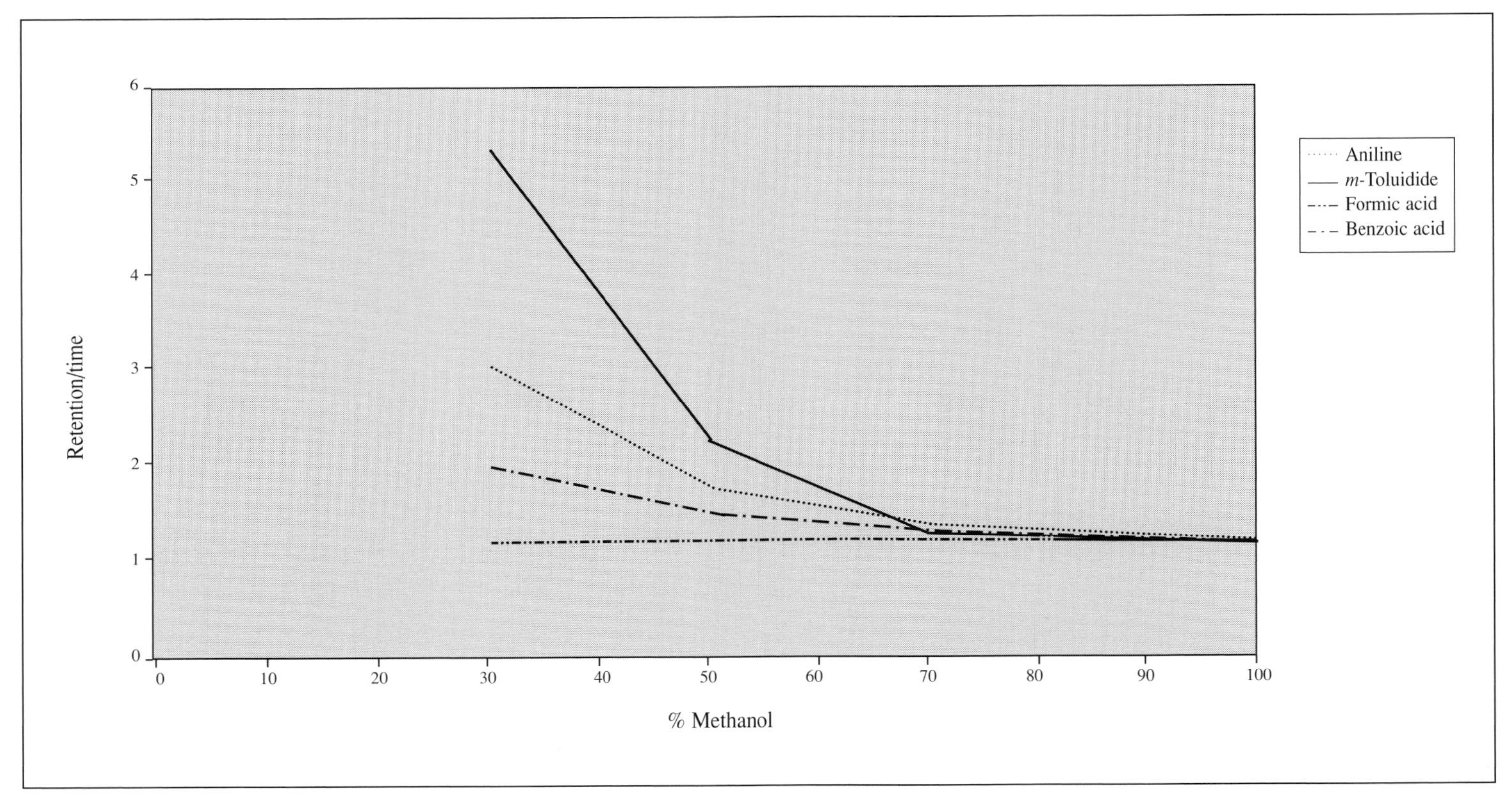

Figure 28-1. RP: dependence of the retention time of organic acids and bases on methanol content of the methanol/water mobile phase – without buffer.

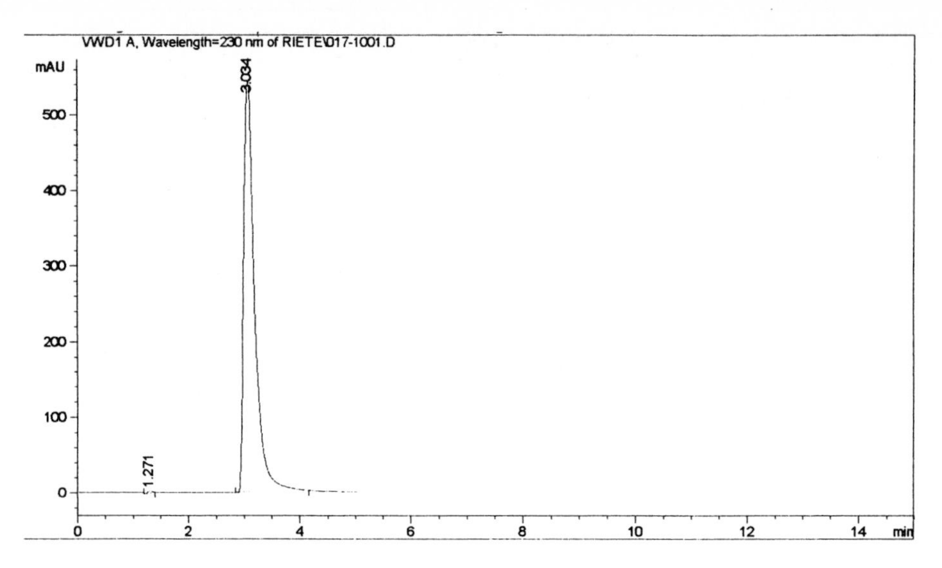

Figure 28-2. Peak form of aniline at 30/70 v/v, methanol/water.

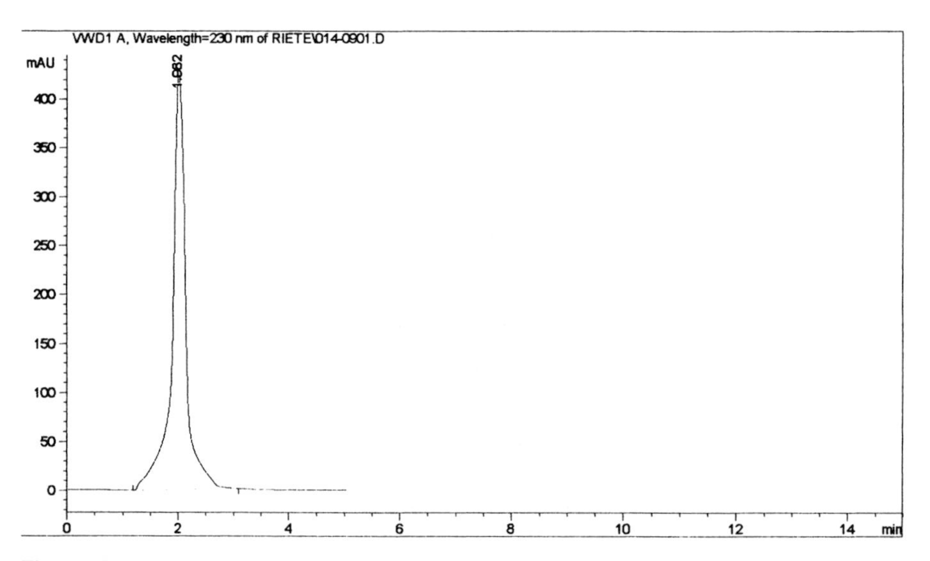

Figure 28-3. Peak form of benzoic acid at 30/70 v/v, methanol/water.

Tip No. 29

What can be the reasons for a change in retention times?

The Case

An unintentional change of the retention time is annoying. The possible causes are in principle the same for all HPLC mechanisms. A changing retention time always involves a change in the following parameters, assuming that column volume (*L*, I.D.) remains constant:

- Δ stationary phase
- Δ mobile phase
- Δ temperature
- Δ flow
- Δ packing density

Also, the procedure to locate and eliminate the problem is similar for all separation mechanisms. However, here we would like to concentrate on the most common mode, reversed-phase chromatography. A systematic list of possible errors can be a significant time saver.

The Solution

We can distinguish three possible causes of the problem:

1. You use the C_{18} column of a different supplier or a column from a different batch of the same supplier.
   - Possibly, the current stationary phase – even though it is still a C_{18}! – has different characteristics from those of the original column, resulting in different retention times with simultaneous changes of *k* values. The "chemistry" has changed (t_m remains constant, t_R changes)
2. You use a new column from the same supplier, same batch.
   - Retention times and t_m increase; the *k* values are constant; the "chemistry" remains constant. The cause of this behavior is a denser packing. The second column contains, for example, 0.9 g stationary phase, the first only 0.8 g.
3. You work with the same column and observe different retention times in the afternoon or on the next day.
   - This is indeed the worst case for routine analysis. We will speak about the most likely causes. Problems observed at extreme pH values are not dealt with.

The first step is a check of t_m. A change of t_m means in this concrete situation a change in the flow. Is this change caused by the pump or by a leakage? The possible causes:

Table 29-1. Most frequent causes of retention time shifts in RPC.

I. t_m changes:

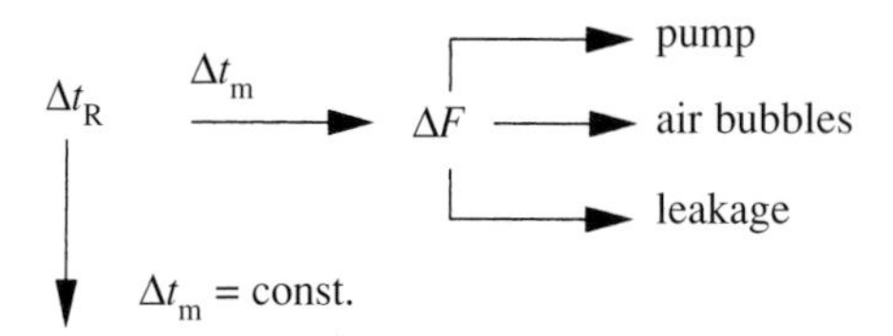

Case 1, 2 or 3 (see below)

II. t_m remains constant:

1. **t_R increases permanently**	→ **One chromatographic parameter changes permanently** • The column is not in equilibrium[2], column saturation effects • Temperature decreases • Enrichment of impurities from the sample matrix
2. **Δt_R, decrease or increase**	→ **There is a fluctuating change in the system** • Occasional air bubbles • Gradient not yet stable? → Longer equilibration • Isocratic? → Mixing OK?
3. **Δt_R, then t_R = constant**	→ **A sudden, singular change in the system** • Has a new mobile phase been used? • pH change[3] • Was it an air bubble? • Temperature change, then constant temperature

Conclusion

If t_m changes at constant column dimensions – and constant packing density! – it is an indication of a flow change within the system. The flow change is either caused by a malfunctioning pump or a leakage in the system.

In order to keep t_R constant, the following conditions should apply:

- constant temperature
- robust mobile phase, constant procedure for mixing the mobile phase, "correct" buffers (see Tip No. 27)
- same conditions of the column, equal time for equilibration, equal start-up procedure.

2) If the mobile phase includes amines or ion pair reagents (see Tip No. 2), equilibrium can take as much as 2 h or more to achieve.

3) A constant pH (correct buffer, sufficient ionic strength) is extremely important for the separation of ions (see Tips Nos. 8, 27).

Tip No. **30**

I use up a lot of RP columns; what should I do?

The Case

You observe that you are going through a lot more reversed-phase columns than you used to and your columns die earlier. Your supplier is not too unhappy, but you would like to know the cause. What does it mean to have a "dead column"? If we exclude higher back pressure, decreased quality means a decrease in the number of theoretical plates (broad peaks, the packing is not OK) and/or a change in retention time (chemical change of the stationary phase). What possible causes should be considered?

The Solution

Unfortunately, there are several possible causes. In the following, we list some of them with comments. Please check whether one or more of the listed chromatographic conditions are part of your current method.

Possible cause	Comments
• Extreme pH of mobile phase	In the alkaline range (pH > 8) dissolution of silica gel; in the acidic range (pH < 2) hydrolysis of C_{18} groups.
• Complex "unappetizing" matrix	In this case, broken particles are possibly better than spherical ones, and 7 mm better than 3 or 5 µm, if your separation does not need high efficiency.
• Temperature higher than 60–70 °C • Water-rich mobile phases (from ca. 80 % water[4]) • Strong (ammonium-containing) buffer, mobile phase with ionic strength of more than 70–100 mM	In all cases, increased dissolution of silica gel. By the way, Li, Na, and K are gentler to the stationary phase than ammonium, and organic buffers such as TRIS in the alkaline range is gentler than inorganic buffers such as phosphate
• Frequent gradient changes	No column likes frequent gradient changes, especially if the gradient starts out with 100 % buffer.

Through a combination of unfavorable conditions, the effect is multiplied. For example, a mobile phase consisting of methanol/70 mM ammonium phosphate buffer pH = 8 (20/80 v/v), is pumped at 55 °C through the column, which is filled with spherical 3 mm material. If you still find after 2 weeks a constant retention time and constant number of theoretical plates, you can invite everybody for a pizza (paid for by the column manufacturer).

4) If you need a water-rich mobile phase think of the newest AQ-phases (eg. YMC AQ, AQUA, Platinum EPS), monolithic media, shield phases, hybrid phases or the old "workhorses" such as Nucleosil 100, Bondapak, LiChrospher

Conclusion

If possible, avoid the above-mentioned chromatographic conditions or give consideration to:

- working with a saturation column (see Tip No. 7)
- possibly using a C_{18} column with a carbon content of ca. 20 %, these being more stable than the more common columns with a carbon content of 7–15 %, but be careful: The capacity will be changed a lot, the selectivity will most likely be changed a little and you always obtain lower plate number.
- possibly use the modern AQ or shield phases (with steric or chemical protection) or phases like XTerra, SilicaROD, Bonus.

Tip No. **31**

Why does my normal-phase system not work any more?

The Case

You work with a "normal-phase system" with silica as the stationary phase and heptane as the mobile phase and everything works perfectly. You prepare a new mobile phase and the retention times are all over the place. How could this happen, since the system worked perfectly well?

The Solution

Poor reproducibility in adsorption chromatography (normal-phase chromatography) is, aside from the limited solubility of polar compounds in organic solvents, the main problem of these systems. Almost always, the problem is with the (uncontrolled) water content in the total system, i.e. the HPLC instrument, the sample, the mobile phase and the column.

Water, which is always present, albeit sometimes just in traces, is adsorbed on the silica surface, and an aqueous layer is formed. The thickness of this layer, which can vary depending on flow, nature of the mobile phase, temperature etc., has a profound influence on the characteristics of polar stationary phases. If the original heptane in our case was "wet", say a water content of 100 ppm, and the freshly prepared heptane is dry, with only 20 ppm water, this difference can have an enormous impact on the separation.

Note. In reversed-phase chromatography, these sudden changes in retention are only seen if the water content changes by a few per cent.

Successful strategy for reproducible NPC results:

Prepare a water-saturated solvent. This is a reproducible procedure. Then you take a really dry solvent – also a reproducible situation – and mix these two solvents to obtain the desired mobile phase. The desired ratio means that ratio which will give a certain water content and hence the desired selectivity.

Conclusion

- In normal-phase chromatography, but also in chromatography with other polar phases (stationary phase: SiO_2 or CN, NO_2, NH_2, mobile phase: heptane, hexane, methylene chloride) a constant water content in the system is *the* fundamental requirement for reproducibility. This also means: take your time for a regeneration of a polar column, a change in mobile phase or simply freshly preparing a mobile phase. Allow some time for the silica surface to form its aqueous layer, at least 2 h, better overnight – or use the above-described short cut; it saves you a lot of time.
- Make sure you have a constant water content in the mobile phase: molecular sieving and Karl Fischer are available to check it.

- Ask the question: "Could the nature of the production process cause the water content of my sample to change?"

And finally:

Maybe this is not the right place to say it, but these problems should not discourage you from trying normal-phase systems out. The selectivity of these systems is often much greater than that of reversed phases, especially if the solutes have the same polarity with small structural differences such as carotenoids, dyes, isomers and phospholipids. Remember: polar groups on the surface might make some trouble but they often have a positive contribution to the selectivity.

Tip No. 32 Chemical tailing at the presence of metal ions

The Case

Chemical tailing is often a problem when separating polar (better: ionic) solutes with HPLC. Free silanol groups on the surface of the silica gel are generally seen as the cause. Today (really for more than 20 years, even if not all manufacturers will admit this ...) we know that metal ions in the stationary phase have a great influence on the separation. Metal ions originate from different sources, such as the manufacturing procedure of the silica gel, the buffers used, stainless steel capillaries, etc. Without going into the problem in more detail, we can make the following assertion: chemical tailing means restricted kinetics. One possible cause is indeed the presence of metal ions which, on the one hand, form loose complexes with ionized solutes and, on the other hand, favor the dissociation of the silanol groups. The result of slow kinetics is broad, tailing peaks. How can you quickly check for metal ions as the source of tailing and how can they be eliminated?

The Solution

(a) Check for metal ions
Inject one of the following compounds:

- 2,3-dihydroxynaphthalene
- 8-quinolinol
- 8-hydroxyquinoline
- acetylacetone
- 2,2'-bipyridyl (according to my experience the most sensitive)

together with a neutral compound, most simply toluene or benzene. If there is a packing defect, you obtain two tailing and/or broad peaks. However, if the tailing is caused by metal ions, the neutral compound elutes as a narrow peak and only the second compound, for example acetylacetone, shows tailing (see Fig. 32-1a).

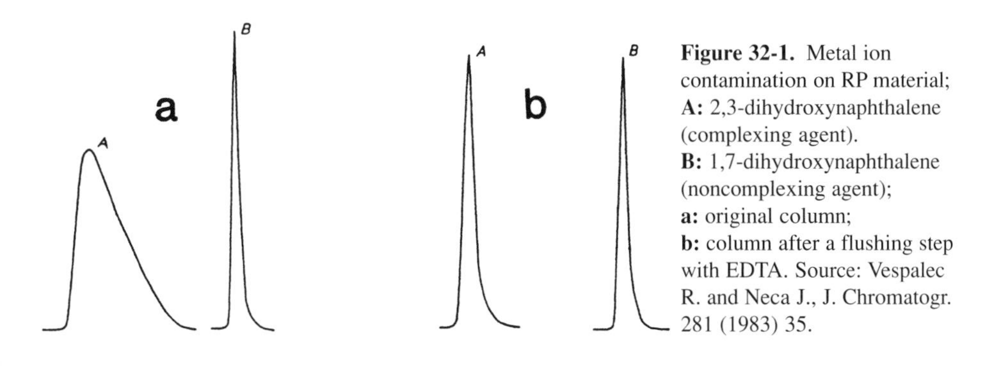

Figure 32-1. Metal ion contamination on RP material; **A:** 2,3-dihydroxynaphthalene (complexing agent). **B:** 1,7-dihydroxynaphthalene (noncomplexing agent); **a:** original column; **b:** column after a flushing step with EDTA. Source: Vespalec R. and Neca J., J. Chromatogr. 281 (1983) 35.

(b) Eliminate metal ions

If you identify metal ions, you can flush the whole instrument with 10–20 mM EDTA or guanidine to complex the metal ions and to flush them out. Another successful flushing solution is a phosphate buffer, pH 2.5. That means, on the other hand, that when working with acid buffers the problem becomes minor. After a successful flushing step, you obtain symmetrical peaks for both compounds (see Fig. 32-1b).

In a large-scale study of NOVIA ("comparison of commercial C_{18} phases"), we compared more than 60 stationary phases using different criteria. Among other tests we looked at the metal ion content of the materials. We found immense differences (see Fig. 32-2). On the (a) you see a metal ion-free phase, in the (b) a contaminated one and on the (c) a very strongly metal ion-contaminated phase. The "disturbance" at ca. 4.7 min is the peak of 2,2'-bipyridyl. There are even phases on the market which are so contaminated that 2,26-bipyridyl does not elute (Fig. 32-3).

Conclusion

Certain complexing agents (see above) are suitable and sensitive indicators for metal ions. If stainless steel and not PEEK is used in the instrument, the metal surface can be one source of the metal ions. The best way to exclude this source is passivation with 6 M HNO_3 followed by flushing with EDTA. From my experience after this comparative study I would also recommend flushing your column with an acidic eluent, e.g. pH 2–3, once a month. A high proportion of the metal ions will be cleaned away. If you only work with complexing agents or with metal-sensitive protein enzymes, an inert HPLC instrument could be an option.

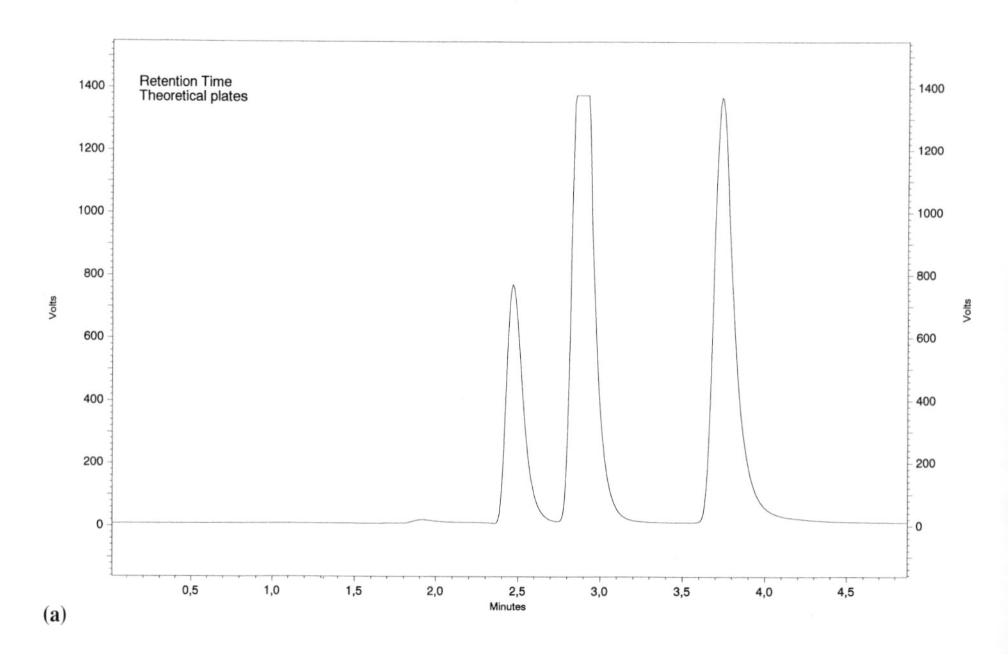

(a)

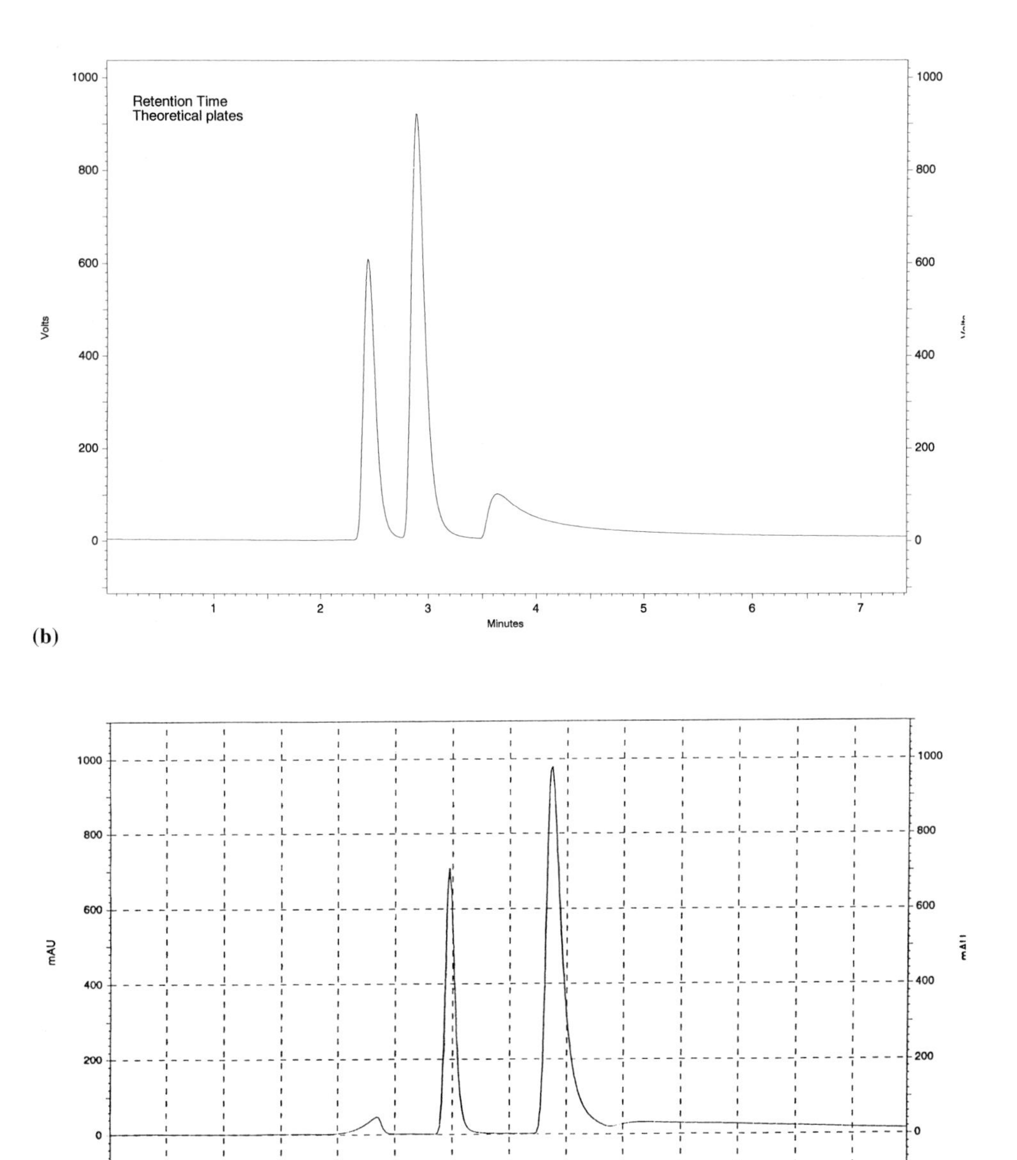

Figure 32-2. Injection of uracil (inert), 4,4'-bipyridyl (polar but not a complexing agent) and 2,2'-bipyridyl (polar, strong complexing agent), eluent 80/20 MeOH/H_2O, v/v at 40 °C, 1 ml/min, 254 nm; stationary phase without metal ions (*left*), stationary phase with contamination of metal ions (*middle*) and stationary phase contaminated with more metal ions (*right*).

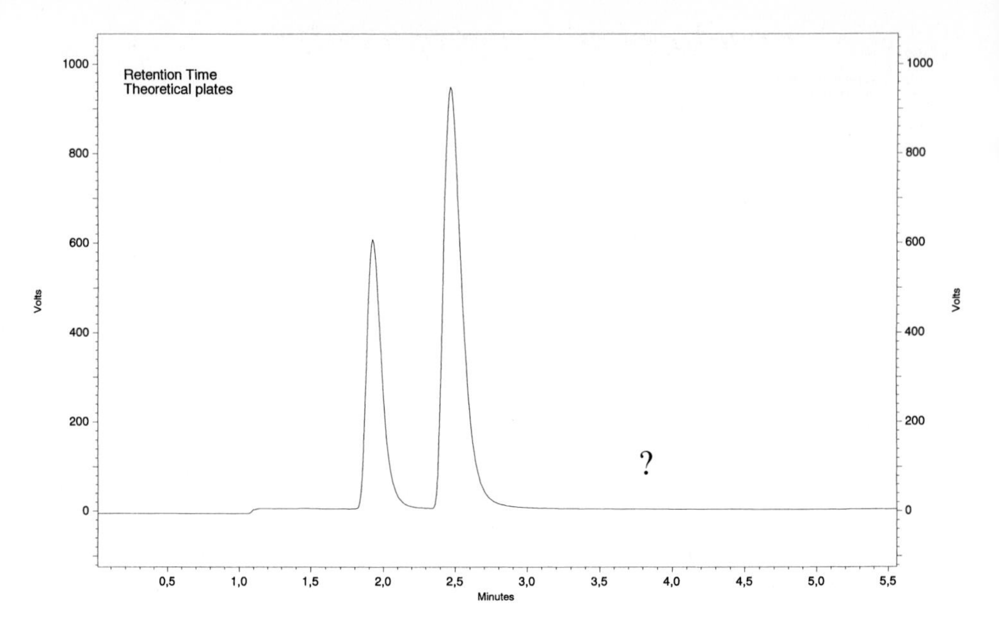

Figure 32-3. Very strong metal ion-contaminated stationary phase, no elution of 2,2'-bipyridyl possible.

Tip No. 33

How to avoid memory effects?

The Case

Unexpected peaks in the chromatogram can have multiple causes: air bubbles, impurities, late elution solutes or memory effects. How do you recognize a memory effect and how can you eliminate it?

The Solution

If a solute is irreversibly adsorbed somewhere in the system and appears later as a peak in the chromatogram as a result of partial desorption, we can speak of a memory effect. The reasons for the partial desorption can be an increased flow rate, a pressure shock, a change of mobile phase or a sudden change of elution strength, e.g. in a step gradient. Alternatively, the capacity of a pseudo-stationary phase could be saturated (see Tip No. 13). It is also possible that a memory effect which has always been present is only now detected because of a change in the wavelength. Compounds are most likely to be adsorbed on injectors, fittings and metal frits.

Peaks originating from memory effect share the following characteristics:

- They always appear at the same time after the injection.
- With multiple injection of the mobile phase, they decrease.

It is simple to check for a memory effect: inject a highly concentrated sample solution, e.g. 5–10 times the regular concentration, and then inject the pure mobile phase. If the signal-to-noise ration is less than 2:1, the memory effect is (if present) negligible.

Recommended Procedures

1. With a constant injection volume (solid loop), inject the loop volume three times, the previous sample solution being flushed through by the current one.
2. Use variable injection volume (only if you have persistent trouble with memory effects, otherwise unnecessary). For example, for a 20 µl loop, first take up 2 µl mobile phase, then the sample. The mobile phase leaves the injection needle last, flushing it.

Should you experience problems with memory effects more often, an improvement in sample preparation and frequent flushing are advised. A very good flushing procedure for the equipment without the column is: isopropanol, water, 6 N HNO_3, flush with water until neutral (it takes some time ...), then again the mobile phase. If you are in a bad mood or don't feel like working hard on Friday afternoon, flush using this procedure. You will surely get a happy, satisfied smile from your equipment. I hope there are 4–5 such Fridays per year in your laboratory.

Conclusion

Non-polar, high-molecular-weight compounds or matrix components are often annoying. They will stick onto rough, "dirty" surfaces. This tells you indirectly what to do. To avoid memory effects, improve your sample preparation and make sure your equipment is clean.

Tip No. 34

How do the default values on my PC affect the resolution?

The Case

The separation happens within the column, but you only can "see" it on the monitor. Using the usual 1-V output of the detector, the settings at the PC are responsible for the optical representation of the chromatogram at the monitor as well as the information. What should you pay attention to?

The Solution

Preliminary remark

If you do not feel competent in the English language, ask your software supplier about manuals in your mother tongue. Some manufacturers will supply these on request; others charge an additional (affordable) sum, and at others you will unfortunately only receive the short version of the comprehensive English manual. It is in each case important to know, how these parameters are to be adjusted and which values the manufacturer recommends for his software.

Now to our subject. Modern software programs offer many options for the manipulation of the chromatogram, e.g. automated baseline subtraction, programmed change of integration parameter, auto-zero functions and auto scale, automated calibration for non-linear signal-to-concentration ratio, etc. It is worth while knowing these options and how to use them as they fit.

The three basic parameters which we will look at in more detail are attenuation, threshold and data acquisition rate (sample rate).

1. Attenuation ("Att")
   Using a 1-V output, the "whole" signal is recorded by the PC. An optimized representation of the chromatogram on the monitor can be obtained by changing the attenuation. This way, you can see more clearly the small peak sitting on the tail of the main peak. A change in the attenuation changes only the optics of a chromatogram; the integration remains unchanged.
2. Threshold, slope
   The software has to recognize a peak as a peak in order to start an integration. That means, the integration starts with a certain slope of the signal height per time unit, *V*/min; this is the threshold. A threshold value that is too small results in a too early start and too late end of the integration. If the threshold is too large, integration starts only at the steeply rising peak slope and ends at the falling part, and the peak area determined is too small.
3. Data rate, data acquisition rate or sample rate
   This simply means the number of recorded data points per time unit. If you record many data points per time unit, the peak is very well represented (too good? waste of too much hard disk space?). If the data rate is too small, the peak is fuzzy, and

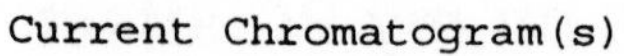
Current Chromatogram(s)

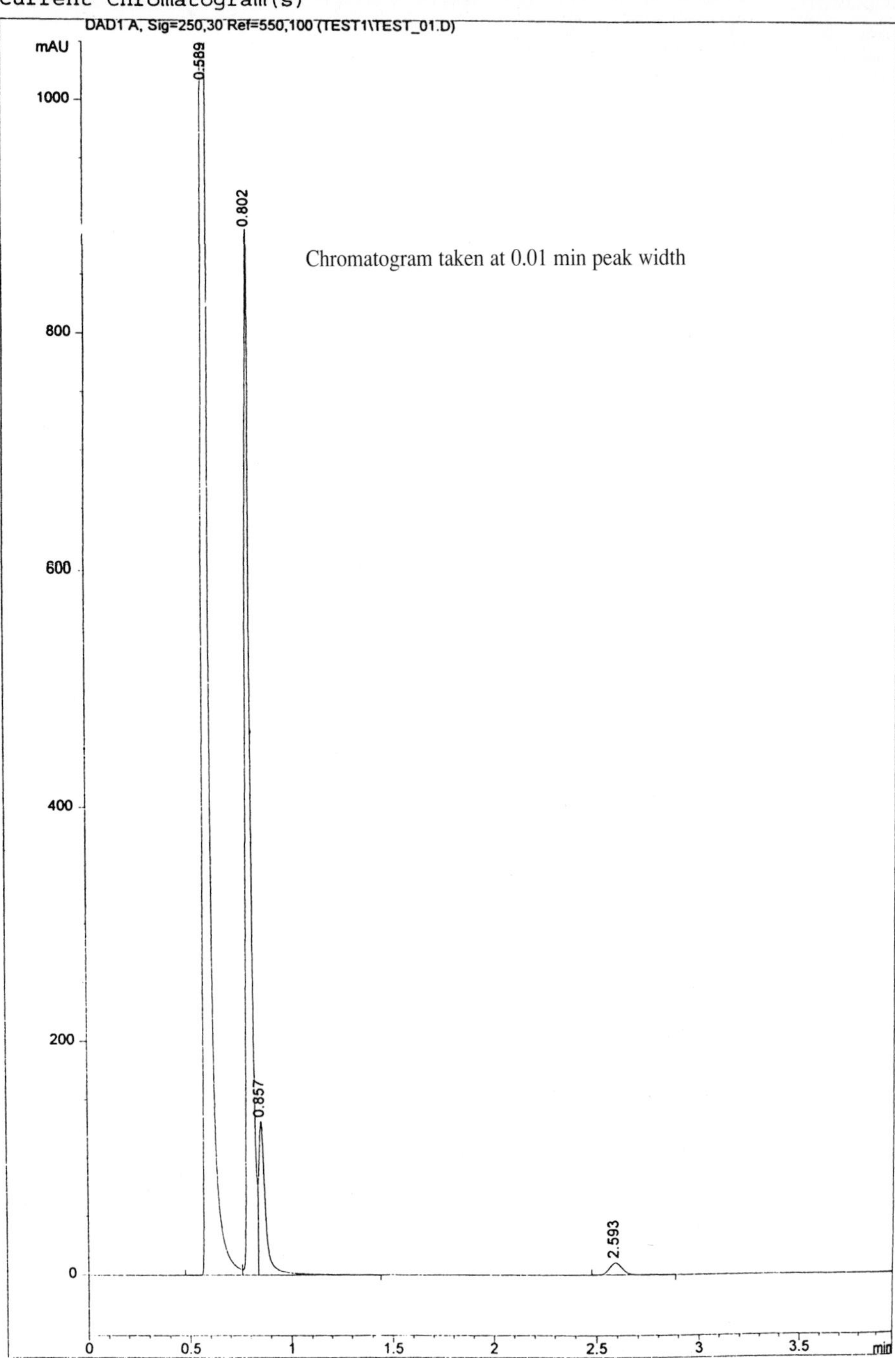

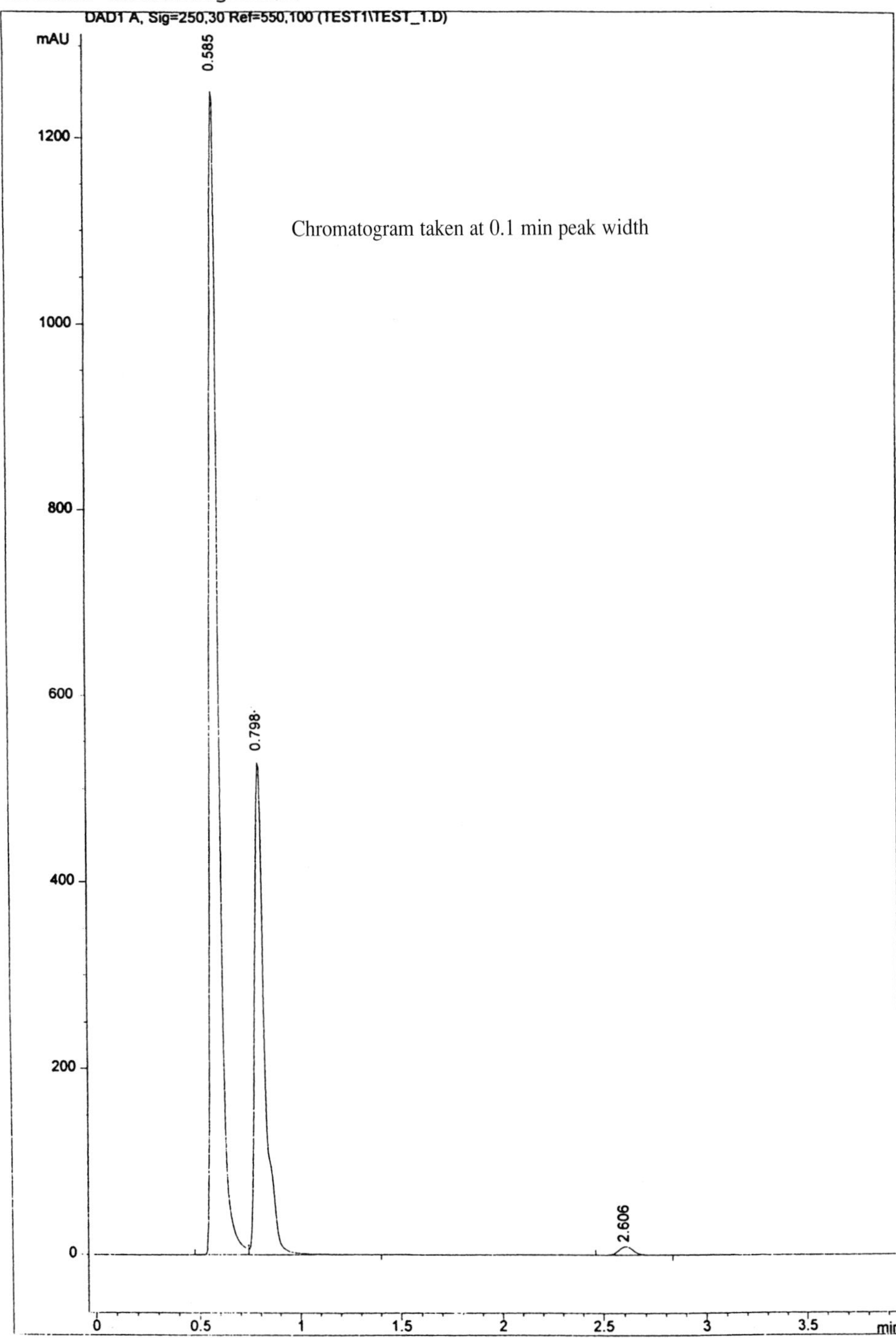
Current Chromatogram(s)
DAD1 A, Sig=250,30 Ref=550,100 (TEST1\TEST_1.D)
mAU
1200
1000
800
600
400
200
0
0.585
0.798
2.606
Chromatogram taken at 0.1 min peak width
0
0.5
1
1.5
2
2.5
3
3.5
min

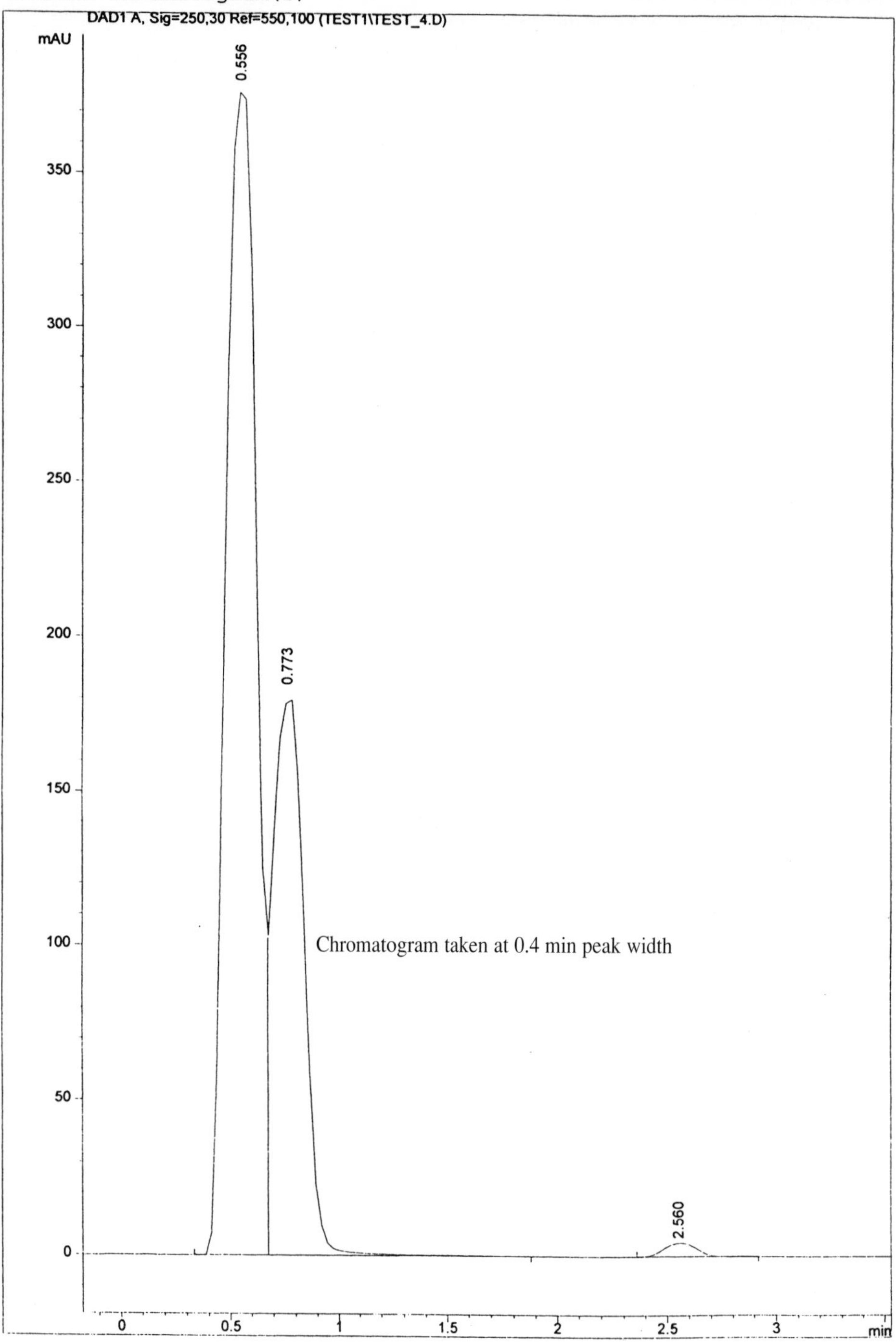

Figure 34-1. Three chromatograms taken at different data acquisition rates, but otherwise under the same conditions (see text).

information can be lost. A peak is fully described by 10–20 data points. A fast peak in "normal" HPLC (20 cm C_{18} column, $k = 1$) has a peak width of 10 s. A data rate of 1 point/s is therefore sufficient; very fast peaks or peaks from micro-bore or short columns or non-porous material should be recorded at 2 data points/s or more. In some software programs, the peak width is linked to the data rate. If the peak width is too large, you lose valuable resolution because of fast and early peaks. For most software programs, peak width only influences the integration, e.g. it decides, based on its size or width, whether a peak should be integrated. Figure 34–1 shows, as an example of the first case, a chromatogram containing three peaks that elute very fast and a fourth one with a "normal" retention time. The left hand chromatogram was recorded with a peak width of 0.6 s, the middle one 6 s and the last one 24 s. At 6 s the third component can only be seen as tailing, and at 24 s the resolution between the first and the second peak is unacceptable. The fourth component at $t_R \approx 2.6$ and $k \approx 4$ simply becomes broader.

Conclusion

If you choose unfavorable threshold and attenuation, you can always change them later; the raw data are saved. Depending on your particular application, please choose an adequate sample rate right from the start. You can only, for example, reintegrate the data that was acquired at the data rate (data points per second) actually used. A repeated injection at as higher data rate means lost time.

4. Tips to Optimize the Separation

Tip No. 35 Which is the right injection technique to get sharper peaks?

The Case

We all wish to obtain sharp peaks in a separation. How to find the best way? We have several options.

One could use more efficient columns resulting in a higher theoretical plate number. In addition, all measures resulting in an improved signal-to-noise ratio (see Tip No. 38) are useful. In this chapter, we will deal with the influence of the injection volume on the peak.

We assume the column is not overloaded and the sample has good solubility.

The Solution

The more or less laminar flow condition in column chromatography causes a Hagen-Poiseuille flow profile, resulting in peak broadening (see Fig. 35-1a). The quest is to limit the development of this flow profile.

The tricks:

- Injection of 0.5 to 1 µl air together with the sample – first the sample, then the air. The air bubble acts as an air cushion, limiting the development of the profile from the injection point to the column. Reaching the column and increasing now the pressure, the air bubble will be dissolved in the mobile phase and will not cause no further disturbance (see Fig. 35-1b).
- Increase the density of the sample solution
  You can inject guanidine or urea with your sample. In this case, the high viscosity of the plug injected in front of the sample limits the development of a Hagen-Poiseuille flow profile. The sample will elute in a smaller peak volume, peak concentration increases and the peak height increases too (see Fig. 35-1b). In a reversed-phase system, urea or guanidine elute with the front and do not disturb the separation.
- Dissolve the sample not in the mobile phase but in a mixture with half the elution strength
  Example: mobile phase 70/30 acetonitrile/water, sample in 30/70 acetonitrile/water. The effect is that the sample is concentrated at the column head, is then transported by the mobile phase and elutes in a small peak volume (see Fig. 35-1c).

Conclusion

Some readers may shy away from a direct injection of air into the system, and increasing the sample viscosity involves a certain effort. The third method is probably the easiest way to obtain small peaks in reversed phase chromatography. The proof is that this approach works and has led to great results with preparative HPLC, where it has been used for some years.

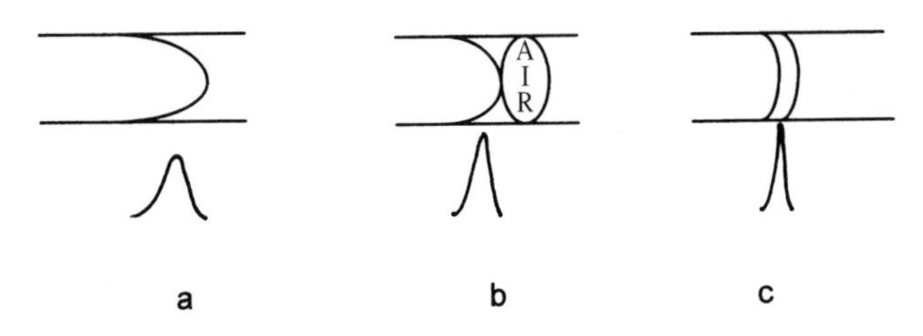

Figure 35-1. Free and forced elution profiles in HPLC.

Tip No.
36

My peaks appear too early – how can I move them in an RP system to later retention times?

The Case

Your peaks elute in a C_{18}, acetonitrile/water or methanol/water system way too early, between t_m and $k = 1$. If your sample is not overloaded, your compounds are most probably very polar or even ionic. How can you increase the retention time?

The Solution

Let us discuss some options:

Case 1: k values stay constant, "chemistry" remains the same, e.g. the interaction with the stationary phase do not change.

Action: Increase column length, lower flow rate.

Case 2: Change of k values, e.g. change of interaction between stationary and mobile phase.

The following three options are available:

(a) Switching to a different stationary phase

Switch to a different C_{18} column (1), same material, but smaller pore diameter (2); do not use endcapped C_{18} phase (3), use C_8-, Phenyl, CN- or diol stationary phase.

(1) A C_{18} column of a different manufacturer can result in unexpectedly large k values, especially for ionic solutes.

(2) Reducing the pore diameter from 300 Å to 100 Å results in an increased specific surface and therefore increased interaction with the stationary phase; the solute will elute later.

(3) The free silanol groups allow ionic interactions, which come in handy in our case, but there is a likelihood of a tailing problem.

(b) Changing the mobile phase

Increase the water content; change the pH (4); decrease the ionic strength (5); use ion pairing reagents (6); see Chapter 5:

(4) Measure pH of the solute:

Acidic? ⇒ Run acidic mobile phase e.g. phosphate buffer at pH 2.5–3.5 or even perchlorate at pH 1.5–2.

Alkaline? ⇒ Optimization more difficult, start at best with 300 ppm TEA or alkaline buffer (see Chapter 5: "Retention of Ionizable Components in Reversed-Phase HPLC".

(5) The mobile phase will be less polar; the solutes are forced to stay longer in the stationary phase and elute later.

(6) For acids, e.g.: 50–100 mM tetrabutylammonium chloride or ammonium hydrogen sulfate, pH ≈ 7.5.
For bases, e.g.: penta- to heptasufonic acid for weak bases and octa- to dodecasulfonic acid for strong bases, pH ≈ 3.5.

(c) Temperature decrease

Conclusion

There are a few relatively simple ways to increase retention times. If they fail, consider another separation mechanism, such as ion exchange or ion exclusion chromatography or move to capillary electrophoresis or capillary electrochromatography.

Tip No. 37

How can I increase the plate number?

The Case

With a high theoretical plate number of a column you can expect sharp peaks. With a given selectivity, resolution is improved by sharp peaks since the resolution is defined as the distance between two peaks at the peak base and not peak maximum. Therefore, you should try to get a high theoretical plate count to obtain a better separation. How can this be done?

The Solution

Note:
Band broadening and tailing can also be the result of column overloading and ionic interaction. In this discussion, we would like to exclude these possibilities. The actions discussed here include the column, the instrument and the chromatographic parameters. Table 37-1 summarizes the options to increase the theoretical plate number.

Table 37-1. Options to increase theoretical plate number.

Option	Comments (if necessary)
Column and stationary phase	
• New column with the same stationary phase, but better packing	
• Smaller particles of the same stationary phase	3 µm are worth it if your isocratic instrument has a dead volume of < 30–40 µl. For gradient separations, the particle diameter is less important.
• Spherical particles	A homogeneous packing is simpler to obtain with spherical than irregular particles. Spherical particles normally have a smaller size distribution compared to irregular particles, also favoring improved plate numbers.
• Longer column	In real life, the disadvantages, such as longer retention times and therefore longer analysis times often outweigh the advantages.
• Non porous particles (NPS)	Not only the theoretical plate numbers changes but also the selectivity, because the stationary phase is different.
• Frequent change of column frits for difficult matrices	Dirt adsorbed on frits often leads to tailing.
Note:	A thinner column will not result in improved theoretical plate numbers but in an increased peak at constant peak area (see Tip No. 38), since the variance in the radial direction decreases.

Option	Comments (if necessary)
Instrumentation	
• Smaller dead volume: If the inner diameter of the connecting capillaries is ≤ 0.2 mm and there is no heat exchanger capillary hidden in the detector and your fittings are OK, your only option to significantly reduce your dead volume is by a smaller detection cell.	
Chromatographic parameters	
• Decrease of flow rate	Using good, homogeneous spherical 3 µm or 5 µm particles, the advantages are often negligible.
• Increase of temperature	For a classical reversed-phase separation, a temperature of 35–40 °C is a reasonable compromise, because a further temperature increase would reduce the selectivity (see Tip No. 16).
• Decrease of the viscosity; acetonitrile/water mixtures better than methanol/water mixtures	Learn something form the experienced HPLC sales representatives, who always run their demo units with acetonitrile/water .... In general an acetonitrile/water, compared to a methanol/water mixture with the same elution strength, results in a chromatogram with almost identical retention times but sharper peaks: acetonitrile is less viscous, kinetics is faster leading to sharper peaks.
• Sample amount	Overloading leads to a decreased theoretical plate number. The more complex the separation mechanism, the greater the danger of local overloading of the column. Remember that even with 20 µl injection volume you could have a local overloading of the column.

Conclusion

A well-packed column with 5 µm spherical silica gel based particles should result in ca. 60 000–80 000 theoretical plates/m. Reasonable efforts of the user to obtain sharper peaks include:

- optimization of the instrumental dead volume
- possible temperature increase
- change column manufacturer – not necessarily the stationary phase if you are happy with the selectivity.

Tip No. 38 Limit of detection: how can I see more?

The Case

A peak looking like this

is certainly not pleasant. How can you decrease the limit of detection, or in other words improve the signal-to-noise ratio?

The Solution

Fortunately, there are a number of relatively easy-to-implement options, resulting at once in a better performance. The simplest ones are marked with a * (see Table 38-1).

Table 38-1. Options to decrease the limit of detection.

	Options	Comments
	Increase of theoretical plate number, the peak will become sharp and high.	Obtain with – better packed column – smaller particles – smaller instrumental dead volume (see Tip No. 37).
*	Detect more sensitively.	Increasing the detector sensitivity by 2 will result in 2 time peak height, however, noise increases only by $\sqrt{2} = 1.4$ (see Fig. 38-1), the signal-to-noise ratio increases. *Note*: This is only effective with a 10 mV output signal. It has been shown that a 10 mV output has general advantages for trace analysis.
*	Is the wavelength setting optimized?	Check!
*	Decrease rise time (response time, time constant).	This is also true for a 10 mV output.
*	Column miniaturization: decreased column length and diameter.	This is one of the most effective ways. The solute is eluted (by a constant injection volume) in a smaller peak volume, the concentration is higher, so that the peak is higher with the same peak area.
	Increase gradient slope.	
*	Optimized integration parameters, for example: sampling rate, peak width?	It is impressive how you can "optimize" a chromatogram with the integration parameters (see Tip No. 34).
	Do you get perhaps an unnecessarily broad peak because of the presence of metal ions?	Please check and eliminate the problem if necessary (see Tip No. 32).

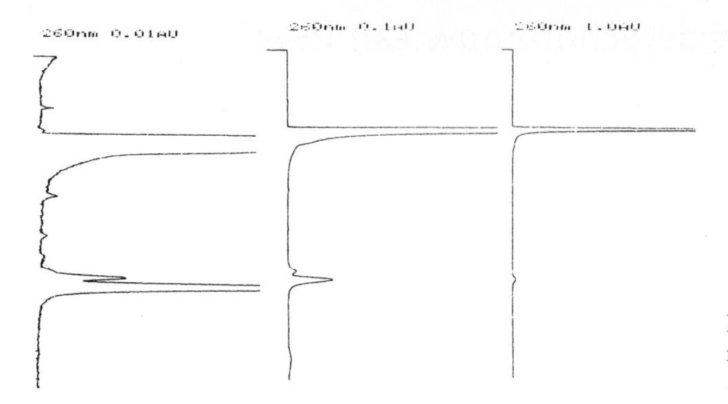

Figure 38-1. Signal-to-noise ratio with different sensitivities: 1 AU, 0,1 AU and 0,01 AU.

Options	Comments
* Inhibit dilution – use simple tricks for injection (see Tip No. 35).	A simple and effective way.
Install a pulse damper after the pump (see Tip No. 25).	The investment of $ 1000 is high for older pumps.
Electrical surge protection or special electronic filter.	The latest ones became available recently. They look good, but there is not enough experience with them for a judgment.
* Increase injection volume or sample concentration.	This is the simplest step.
* Are detection cell and mirror clean?	Flush with isopropanol or hot water (see Tip No. 21).
Did you degas sufficiently?	See Tip No. 5.
If a low detection limit is a persistent problem, you should look for a detection cell with an improved design.	With an improved design of the UV cell, the light yield is better and the signal-to-noise ratio improves.
Is derivatization an alternative?	
Use non-porous resins.	

Conclusion

If you occasionally or constantly have to perform trace analysis, optimal chromatographic parameter are more important for you than for “normal” users. In addition, the subject of “miniaturization” in HPLC should be your constant companion.

Tip No.
39

How can I speed up a separation?

The Case

Let us assume you have succeeded in obtaining a good separation. Maybe the resolution is even too good, which would be shown by the unnecessarily large distances between the peaks of interest or the unnecessarily long time between injection and the elution of the first peak. The next reasonable step would be to obtain a sufficient separation with shorter retention times. What are the options?

The Solution

Enclosed are a list with some simple (•) and some more elaborate options

Table 39-1. Options to decrease the retention times.

	Shortening the retention time by using one or more of the methods listed below
	A shorter column; a 10 cm column with 3 mm particles delivers a similar separation to that of a 15 cm column with 5 μm particles – in about 70 % of the analysis time.
	A thinner column by constant flow rate
	Same stationary phase, but with particles with larger pore diameter (see Tip No. 36)
	Instead of C_{18} take C_8. An interesting option for creative users, who like to work with precolumns: precolumn C_{18}, analytical column C_8, or precolumn C_8 and short analytical C_{18} column, or ...
	Non-porous particles. Problem: expensive, high back pressure, low loadability, high requirements of the equipment
•	Increased flow rate or gradient slope
•	For gradient runs: steeper slope and/or reduction of gradient volume (see Tip No. 45)
•	Temperature increase (see Tip No. 16)
•	Increased content of organic modifier in mobile phase
	For ionic solutes: change in pH, increase of ionic strength
	Tertiary mobile phase: For example, replace 50 methanol / 50 water with 40 methanol / 10 acetonitrile / 50 water.

Conclusion

If you do not do preparative work, the goal should be not the maximal resolution, but the optimal or sufficient resolution. If you have a selective chromatographic system, you only need a small theoretical plate number for your separation. You can even afford to give away plates by increasing the temperature if you gain on time (see Tip No. 16).

Tip No. **40**

How can I optimize a separation?

The Case

The question is deliberately vague in order to make a point. Before you optimize a separation, you have to accurately define what you want to optimize. For analytical HPLC, optimization often means:

- increasing the distance between peaks, i.e. increasing the resolution – *separate better*
- decreasing the retention times ⇒ *separate faster*
- decreasing the detection limit ⇒ *see more.*

Tips Nos. 38 and 39 deal with the decrease of the detection limit and the retention time. In this question, we talk about the improvement of the resolution. We assume that your equipment is optimal. In general, think small, i.e. think miniaturization. Now, how can I separate better?

The Solution

First let us review the parameters influencing the resolution (see also Section 1.4). $R = \mathrm{f}(k, \alpha, N)$

Parameter	**Meaning**
Resolution, R	Distance between two peaks at the peak base (simplified)
Capacity, k	Degree of interaction of a solute with the stationary phase in a given chromatographic system.
Selectivity, α	Capability of the chromatographic system to discriminate between two or more solutes.
Efficiency, N	Measure of the broadening of a solute band in the equipment, essentially in the column.

To improve efficiency, see Tip No. 37.

A change in capacity/selectivity means a change in chemistry/thermodynamics, e.g. a change of stationary/mobile phase and/or temperature. A change of the capacity can be obtained by relative simple means (see Tip No. 36). However, a deliberate change in the selectivity is much harder to achieve. Luck, experience and feeling are part of it.

Table 40-1. Options to change capacity and selectivity (RP).

Chromatographic parameter	Factor influencing capacity/selectivity
1. Stationary phase	– Matrix (base material) • Matrix nature: silica gel, aluminum oxide, titanium oxide, polymer, graphite etc. • Characteristics of the matrix: specific surface, pore diameter, impurities, porous/non-porous particles etc. • pretreatment: acid wash, pore extension etc. – Modification (for reversed phase) • Nature of the functional group: C_{18}, C_8, CN, NH_2, ... • Method of binding alkyl silane: mono-, di, trifunctional. • Carbon content • Special modification: endcapping ("classical" or hydrophilic), introduction of protecting group (steric or chemical protection), covering surface with complexing reagents, change of surface structure, etc.
2. Mobile phase	• Elution strength of mobile phase • Modifier • Important for polar solutes: pH, nature of buffer and buffer concentration, concentration of PIC reagents
3. Temperature	A decrease in temperature is accompanied with always an increase in capacity and in most cases an increase in selectivity.

Conclusion

You will have to define what to optimize and possibly set your priorities. The optimization should occur in the sequence: capacity, selectivity, efficiency (see Tip No. 41).

In real life, the following sequence for the optimization of resolution works out fine and is relatively fast, assuming optimal sample preparation:

- Change of mobile phase, for ionic solutes including pH, buffer and ionic strength
- Decrease or change of temperature
- Change of stationary phase or use of columns in series.

The following steps mean more effort, but result in better information:

- Off-line/on-line coupling with another chromatographic mode (TLC, GC) or related techniques (CE, CEC)
- Off-line/on-line coupling with a spectroscopic method e.g. MS(MS) MALDI, NMR.

Now, let us have a look at the most important ways of getting information about the nature of a sample:

1. Very good but laborious:
   Coupling of two (separation) techniques (hyphenated or orthogonal techniques, multidimensional chromatography) e.g. HPLC-GC, HPLC-TLC, HPLC-CE, HPLC-Elisa, HPLC-Westernblott.
   Remember that after coupling two separation techniques we have a multiplication of the peak capacities of the two techniques, e.g.: You may separate 10 compounds with an HPLC method and 10 with a TLC method, if you couple this HPLC method with the TLC method you could theoretically separate 100 compounds!

2. Very good, fast, the best – if it works
   Coupling of chromatography and spectroscopy ("chromatoscopy") e.g. HPLC-MS(MS), HPLC-FTIR, HPLC-NMR, HPLC-NMR-MS.
3. The best but the most expensive and laborous.
   Combination of 1 and 2, which means coupling of two different separation techniques followed by a coupling with spectroscopy, e.g. HPLC-GC-MS, HPLC-CE-MS.
   The future belongs to miniaturization and the use of specific methods (non-selective as chromatography is). Chip technology and several kinds of spectroscopy are coming. We will see the shift of "chemistry" (separation) to "physics" (determination). For example, in the coupling LC-MS(MS) the LC term will become more and more unimportant. MS(MS) will be one of the winners in the coming years. In general, future winners will be those approaches with the characteristics "robustness" and "specificity", as these guarantee trustworthy information in a short time, e.g. NMR, sensors, molecularly imprinted surfaces. For more details, see "Trends in HPLC" in the appendix.

Tip No. 41 Dead volume, capacity factor, selectivity – how can I use them in everyday life?

The Case

As in almost every HPLC book, the author of the present book also describes theoretical concepts (see Section 1.5 of this book "Some Important Chromatographic Concepts"). Is this because each author feels obliged to do this or is there something that a down-to-earth practitioner can gain from the theory?

The Solution

Yes, most definitely.

The sequence of rational steps for optimization in HPLC is:

1. adjust the k range to the optimal value
2. improve selectivity
3. increase efficiency.

A chromatogram can tell you often, in which state of optimization you are. In Table 41-1 we assume a reversed phase system. In the left column, you will find example chromatograms with one solute.

Table 41-1. Chromatograms, parameters and the next optimization step.

Relevant chromatograms/parameters	Insight/advice on subsequent steps
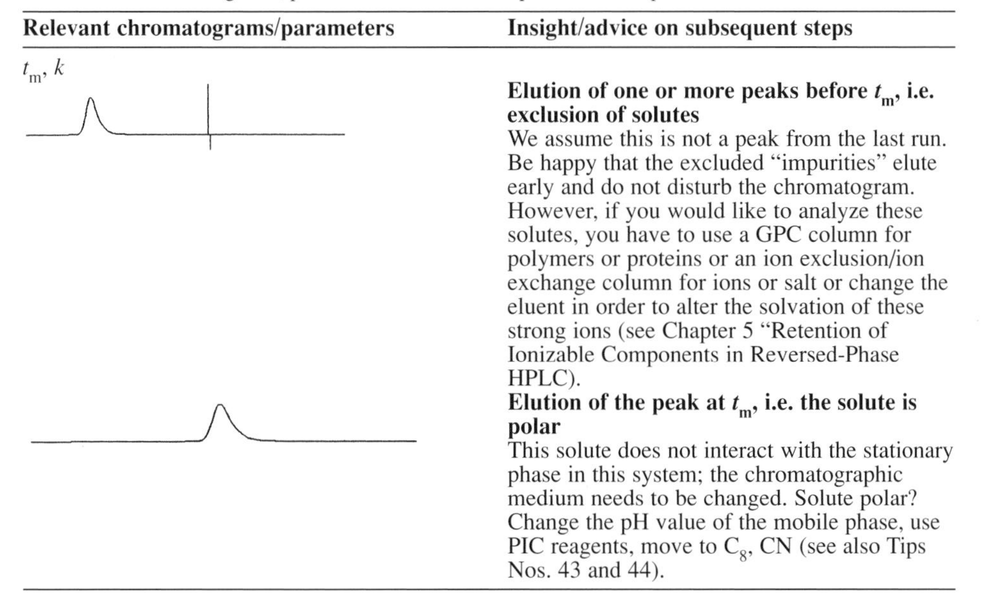 t_m, k	**Elution of one or more peaks before t_m, i.e. exclusion of solutes** We assume this is not a peak from the last run. Be happy that the excluded "impurities" elute early and do not disturb the chromatogram. However, if you would like to analyze these solutes, you have to use a GPC column for polymers or proteins or an ion exclusion/ion exchange column for ions or salt or change the eluent in order to alter the solvation of these strong ions (see Chapter 5 "Retention of Ionizable Components in Reversed-Phase HPLC).
	Elution of the peak at t_m, i.e. the solute is polar This solute does not interact with the stationary phase in this system; the chromatographic medium needs to be changed. Solute polar? Change the pH value of the mobile phase, use PIC reagents, move to C_8, CN (see also Tips Nos. 43 and 44).

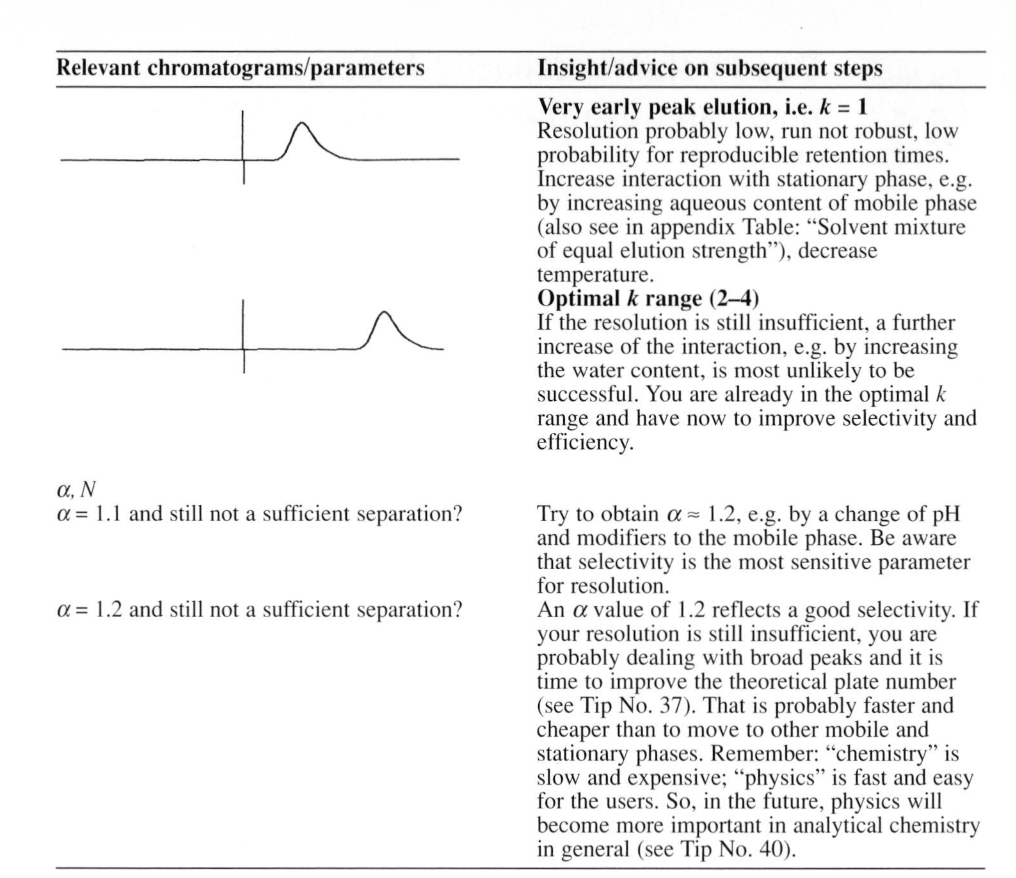

Relevant chromatograms/parameters	Insight/advice on subsequent steps
	Very early peak elution, i.e. k = 1 Resolution probably low, run not robust, low probability for reproducible retention times. Increase interaction with stationary phase, e.g. by increasing aqueous content of mobile phase (also see in appendix Table: "Solvent mixture of equal elution strength"), decrease temperature.
	Optimal k range (2–4) If the resolution is still insufficient, a further increase of the interaction, e.g. by increasing the water content, is most unlikely to be successful. You are already in the optimal k range and have now to improve selectivity and efficiency.
α, N	
$\alpha = 1.1$ and still not a sufficient separation?	Try to obtain $\alpha \approx 1.2$, e.g. by a change of pH and modifiers to the mobile phase. Be aware that selectivity is the most sensitive parameter for resolution.
$\alpha = 1.2$ and still not a sufficient separation?	An α value of 1.2 reflects a good selectivity. If your resolution is still insufficient, you are probably dealing with broad peaks and it is time to improve the theoretical plate number (see Tip No. 37). That is probably faster and cheaper than to move to other mobile and stationary phases. Remember: "chemistry" is slow and expensive; "physics" is fast and easy for the users. So, in the future, physics will become more important in analytical chemistry in general (see Tip No. 40).

Conclusion

We all are short of time. The tools discussed above are valuable. A thoughtful inspection of a chromatogram is more productive than the simple "try something" approach. The position of your peak relative to t_m tells you what the next optimization step should be. Only if an improvement in the selectivity would not be justified by the expected effort needed to achieve it would one go for the last resort – an increase of the theoretical plate number.

Tip No. 42

Which flow is optimal for me?

The Case

In HPLC, the flow rate has a profound influence on the separation, because retention times and separation performance (theoretical plate number, efficiency) of a column depend on it. A decreased flow rate results in later elution of the peaks and in increased distance between them and vice versa. The relationship between flow and column efficiency is described by the Van Deemter curve (see Fig. 42-1).

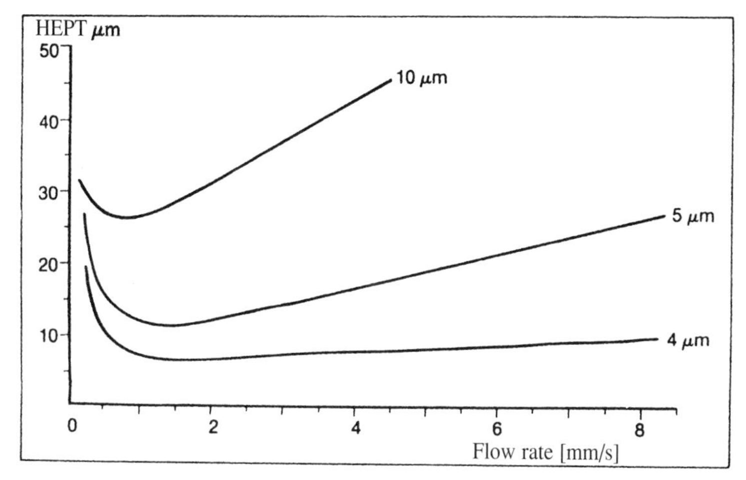

Figure 42-1. The dependence of the height of equivalent theoretical plate (HETP) on the flow velocity of the mobile phase for LiChrospher RP® phases with different particle diameters. The test solute used was benzo(a)anthracene and the mobile phase was acetonitrile/water 75/25 (v/v) (Source: Merck).

The flow at the minimum of the Van Deemter curve corresponds to the optimal flow velocity (better: linear velocity, mm/s), because at this point the highest plate number is obtained. Which flow rate is optimal and how difficult is it to keep it there?

The Solution

Using lower flow rate, efficiency is in general better, but the Van Deemter curve is non-uniform. It depends essentially on the particle size, the nature of the stationary phase, the structure of the solute and the temperature. For small particle sizes (3–5 µm) the minimum is relatively flat and the slope of the Van Deemter curve is not steep. In reality, this means, that with 3–5 µm columns it is possible to work at higher flow rates – pressure permitting. The chromatographic analysis is faster and you do not lose a lot in separation performance. A decreased flow rate would not lead to a noticeable improvement of the separation. (A note in parenthesis for those interested: improving the resolution by decreasing the flow rate becomes more effective as the *C* term of the Van Deemter equation increases. A large *C* term means a low mass transfer rate caused by a slow kinetics for some separation mechanisms).

In the following examples, I recommend to check if you can optimize the separation at a low flow rate (ca. 0.8 ml/min). However, you will have to take the increased retention time into account:

- when using 7 μm or 10 μm particles, especially with irregular particles
- if the column temperatures are < 25 °C.
- using polymeric or polymer-coated stationary phases or phases with double endcapping and/or high carbon content such as Inertsil ODS2 (18 % C), Purospher (18.5 % C), Kromasil (19 % C) or Nucleosil HD (20 % C).
- using a mobile phase with high viscosity such as methanol/water 40/60
- for complex separation mechanisms with a slow kinetics, large and/or solvated molecules, dual adsorption mechanism, etc., for example
  - chiral separations with ligand exchange
  - ion exchange
  - affinity chromatography.

Conclusion

Decreasing the flow rate is a fast way for optimizing the separation. The positive effect is however small for day-to-day routine analysis (C_{18} column, 5 μm or 3 μm particles, acetonitrile/water mobile phase).

Tip No. **43**

How can I optimize a gradient elution?

The Case

A gradient separation is most effectively optimized by means of the gradient profile, the slope of the gradient and the gradient volume. A desired resolution in a short analysis time is often only possible by using non-linear gradients. Therefore, also keep convex, concave or composite linear gradients with isocratic steps in mind. To optimize a gradient, commercial software packages for that purpose such as DryLab and ChromSword are extremely helpful. The second option for an optimization is to change the gradient volume. Let us take a closer look at this second option, because it can be quickly put into operation without need of a commercial optimization program.

The Solution

1. Improvement of the resolution

The peak capacity (number of separated peaks) increases with the gradient volume.

$V_{Gr} = t_{Gr} \cdot F$

where V_{Gr} = gradient volume
t_{Gr} = gradient time
F = flow rate.

If you wish to obtain a better resolution, you will need a larger gradient volume, either by increasing the gradient time, e.g. a shallower gradient at constant flow rate, or, more elegantly, by increasing the flow rate at constant gradient time.

Numerical example

A separation with a flow (F) of 1 ml/min is achieved with a gradient time of t_{gr} = 20 min. With these conditions, the resolution is insufficient and needs to be improved.

The gradient volume computes to V_{Gr} = 1 ml/min · 20 min = **20 ml.** V_{Gr} should increase.

Option 1: $V_{Gr(new)}$ = 1 ml/min · 30 min = **30 ml**
Improvement of the separation with a longer gradient time

Option 2: $V_{Gr(new)}$ = 1.5 ml/min · 20 min = **30 ml**

Improvement of the separation due to an increased flow rate at the same overall analysis time as before (20 min)!

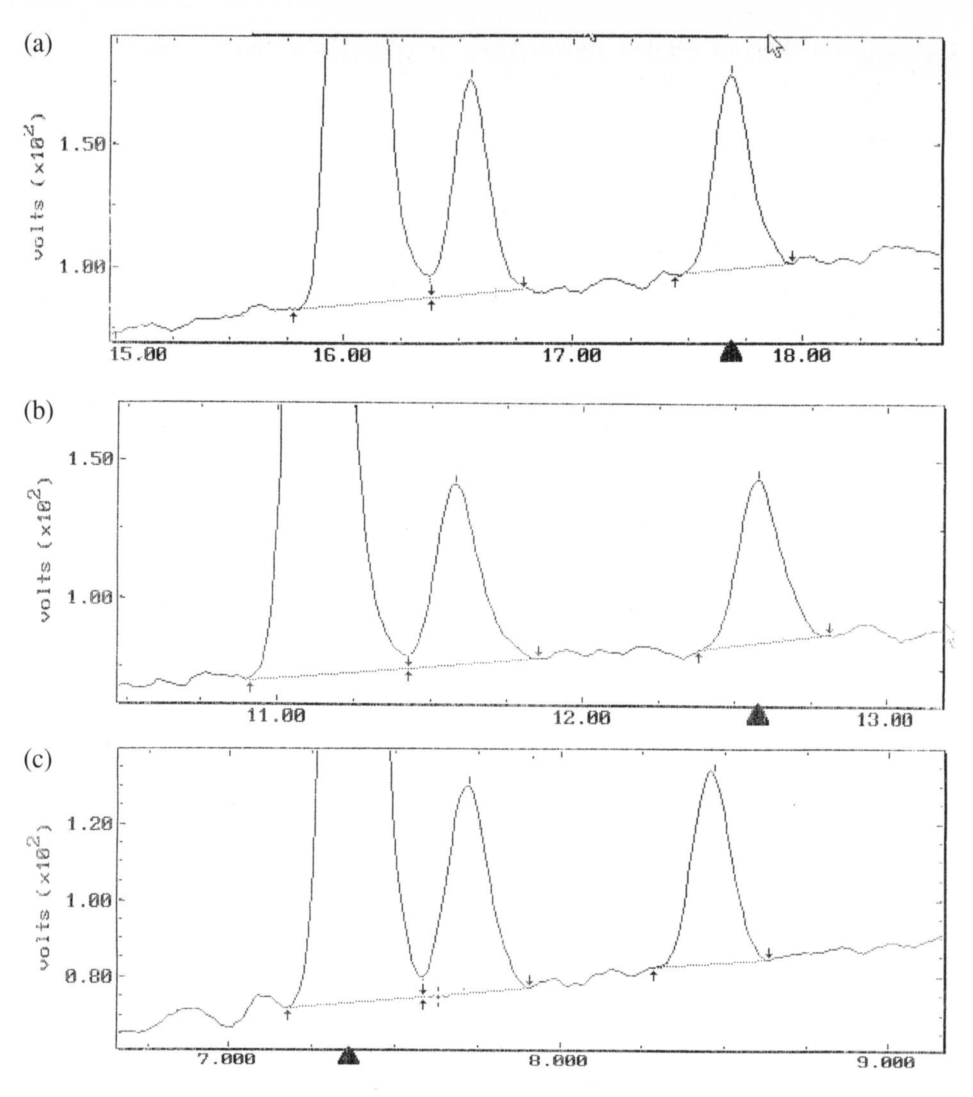

Figure 43-1. Gradient volume (part) of polyaromatic hydrocarbon at (a) 0.5 ml/min; (b) 0.8 ml/min and (c) 1 ml/min.

2. Time Savings

Let us assume you are happy with the separation at the conditions $F = 1$ ml/min and $t_{Gr} = 20$. By doubling the flow and at the same time shortening the gradient time by a factor of 2, you could save time without compromising your resolution. The gradient volume is constant (20 ml), but your separation is finished in half the time!

$$V_{Gr(new)} = 2 \text{ ml/min} \cdot 10 \text{ min} = 20 \text{ ml.}$$

Figures 43-1 demonstrate an example of this time saving by showing part of a separation of polyaromatic hydrocarbons (PAHs). As we go from 0.5 ml/min to 1 ml/min the time required changes from 18 min to < 9 min, an enormous time saving achieved at the same resolution. In all three cases, the gradient volume is constant at 9–10 ml (0.5 ml/min · 18 min = 9 ml and 1 ml/min · 9 min = 9 ml).

Conclusion

In gradient runs, try to work with the highest possible flow rate. Either you gain resolution because of the greater gradient volume or you gain time with the same gradient volume and hence resolution. In a few cases a resolution decrease of 10 % was reported as the flow rate was increased, but the value of the time savings outweighed this disadvantage. As well as the gradient volume approach the following possible to changes can be made in order to get a better gradient separation.

- Lower percentage of B: e.g. instead of 40 % water, start with 10 %
- Use concave or convex profiles and/or enclose isocratic steps in your gradient profile
- Change the slope of the gradient
- Combine a “normal” gradient (increase in the eluent strength) with a flow gradient in one step. Use short columns for this, or use monolithic media.

Tip No. 44

Separation of ionic solutes: what works out best – endcapped phases, inert phases, phosphate buffer or ion pairing reagents? Part I

The Case

A few, several or all peaks in your chromatogram elute with tailing. There are various possible causes, such as column overloading, not optimal parameter setting on the detector or data processing unit (rise time, data collection rate), too large a dead volume in the system. Or maybe is it related to the column? Assuming you can exclude the first three causes and the dead volume is also within the limits, then there is only the column. The question is: what exactly is wrong with the column? How can you check it and what are your options to correct it?

The Solution

If the above causes can be excluded, two possible reasons for tailing remain:

1. The packing quality has deteriorated. Cracks, channels and hollow spaces are present.
2. You injected ionic solutes, interacting with ionic parts of the stationary phase (residual silanol groups, metal ions). The slow adsorption kinetics and especially the slow desorption results in chemical tailing.

In an actual case, what is the cause of the tailing? Is it chemical tailing or is it due to the packing quality?

In order to find out, you inject a neutral solute: toluene, ethylbenzene, acetophenone or something similar. If the packing is defective, you always obtain tailing peaks, independently of your injected solute. If the column is defective, it is defective for all solutes. In this case, you replace it with a new column. If, however, the injection with toluene results in a reasonable peak, you know what the problem is: additional, unwanted interactions of your solute with the stationary phase.

Measures

Preliminary note It is always worthwhile to measure the pH of your sample solution. This information can be very helpful for the rational choice of the stationary and mobile phases.

1. *Mobile phase*
   - Adjust the pH with a buffer to ensure that the solutes in question are not dissociated.
   - For the separation of bases, the residual silanol groups can be blocked by addition of triethylamine or octylamine in ‰ amounts. For the separation of acidic components, the addition of 0.5–1 % acetate, phosphoric acid or

perchloric acid will suppress the dissociation of the sample components and the silanol groups.
- Increase ionic strength.
- For strong acids or bases: add ion pair reagents

2. *Column*
   - Depending on the sample characteristics (acid/base), use a column with the "right" silica gel (see Tip No. 3)
   - Use "endcapped" (silanized) stationary phases, which can often be recognized from trade names of the form XYZ **e,** XYZ **E**, XYZ **II**, or can be referred to as deactivated or shield columns.
   - Use metal-free stationary phases, available under different names in the market place (see Tip No. 1), e.g. Inertsil, Kromasil, Luna, Zorbax SB, Symmetry, Purospher ....
   - Complex and eliminate metal ions with EDTA

Conclusion

The separation of ionic solutes should be accomplished without ionic interaction. The special case that ionic interactions are purposely used for selectivity improvement is excluded here.

Column
Choose metal-free, endcapped, deactivated, shielded stationary phases for bases and non-endcapped for strong acids.

Mobile phase
Reduce the influence of the silanol groups through additives in the mobile phase and/or the pH or with ion pairing reagents. Take care that the solutes in the mobile phase exist as neutral, not dissociated molecules.

Tip No. 45

Separation of ionic solutes: what works out best – endcapped phases, inert phases, phosphate buffer or ion-pairing reagents? Part II (see also Chapter 5)

The Case

In Tip No. 44, we started to deal with the most difficult problem in reversed-phase chromatography, the separation of polar and ionic organic solutes. In this Tip, the subject is further elaborated. At least if you obtain tailing peaks and have excluded all other possibilities (see Tip No. 44) you know you deal with ionic solutes, where bases are more annoying than acids. The conditions suitable for the separation of ionic solutes are listed below in Table 45-1. Stationary phases and mobile phases listed side by side in the Table do not necessarily have to be used together; it is possible to succeed with one or the other.

The Solution

Table 45-1. Recommended chromatographic conditions for the separation of ionic solutes.

Sample	Stationary phase	Mobile phase
Polar solutes		
	Deactivated, polar RP phases and/or phases with a high carbon load.	Water-rich mobile phases.
Very polar solutes	Very well endcapped or double endcapped C_{18} phases or C_8-, C_4-, CN-, diol phases (also for bases, see above).	Very water-rich mobile phases, for example > 95 % water.
Bases	Metal ion-free, modern stationary phases suitable for separating bases; these include Symmetry, Kromasil, Inertsil, Nucleosil AB, Purospher, Zorbax SB, SymmetryShield, HyPurity Elite, Prodigy, XTerra, Bonus (think of C_8 too!).	First choice: acidic pH (e.g. pH 2.5–3.5) with acetonitrile; if retention too little, move to pH 7.5 with methanol as the organic modifier. You may also add triethylamine or octylamine (ca. 50–500 ppm) or for strong bases add an ion pair reagent (PIC B), e.g. 50–100 mM octasulfonic acid.
Acids	The column is less critical compared to bases, a good (not endcapped) C_{18} would be adequate *Note*: Many stationary phases specially produced for bases are not suitable for use with acids.	Acetate, perchloric acid, phosphoric acid or phosphate buffer pH 2–3, PIC A reagents.
Acids and bases	Endcapped C_{18} phases.	Acetate and triethylamine or octylamine.

Conclusion

A simplified scheme for solving separation problems with ionic solutes is:

- For the separation of bases: Use of modern C_{18} or polar RP stationary phases with very acidic mobile phases; alternatively with the opposite, namely alkaline buffers, if necessary with PIC B reagents in the mobile phases.
- For the separation of acids: Low pH of the mobile phase, if necessary containing PIC A; classical stationary phases, not endcapped.

Ionic solutes can be handled with the help of modern stationary phases and – if necessary – the use of additives in the mobile phase. If this should not work, you can as a last resort turn to stationary phases based on polymers, aluminum oxide, titanium oxide, graphite etc. Or you can make use of an ionic separation mechanism such as ion exchange, or forget HPLC and think of CE or CEC, or just forget everything and go to see the next Star Wars episode.

About ionizable solutes, sun, and lectures in the afternoon ...

Dear reader, I think we are in agreement that the separation of ionizable organic solutes is one of the most difficult problems in RP-HPLC (Tips Nos. 27, 28, 44, 45). Of course, we all know that in such systems the pH and the ionic strength are very important and that we have to contend with tailing and so on. For years now I have looked in vain for a nice, plain, easy-to-understand essay about these things. That is the background.

Then in June 1999 there was the 23rd International Symposium on High-Performance Liquid Phase Separations and Related Techniques (HPLC '99) in Granada, Spain. A poster of the separation of ionizable solutes positively astonished me. One of the authors, Mr LoBrutto, was to give his lecture – the last one of that day – at the impossible time of 5.50 p.m.!

It was a wonderful day, the sun was shining over the Alhambra, my wife was waiting for me in my hotel and this Mr LoBrutto's turn was at 5.50 p.m. I was rather furious but I stayed. It was worth it.

Briefly, Rosario LoBrutto and Yuri Kazakevich had earlier sent me some ideas. We talked about these, among other topics, and the result was the following chapter about important factors that influence the retention of ionizable components in RP-HPLC.

Enjoy it!

5. Retention of Ionizable Components in Reversed-Phase HPLC

R. LoBrutto, Y. Kazakevich
Seton Hall University, Chemistry Department
400 S. Orange Ave. S. Orange, NJ 07079

5.1 Introduction

5.1.1 History

HPLC separation of ionic or ionizable components was first attributed to ion-exchange mechanisms [1–3]. In this process the retention of ionic analytes is governed by their ionic interactions with ion-exchange sites embedded in the packing material [4, 5]. The process appears to be very inflexible for the separation of organic ionizable compounds, which are usually weak acids or bases. Tools for the adjustment of the selectivity of separation are very limited in this mode. The separation of closely related organic bases or acids with small differences in chemical structure are almost impossible in an ion-exchange mode.

A further modification of the ion-exchange process was the introduction of ion-pair chromatography. This mode is a hybrid of the reversed-phase and ion-exchange processes. The addition of an ion-pairing agent (some form of surfactant) into the mobile phase causes the formation of ion pairs with ionic analyte molecules. It is believed that these neutral pairs are retained in the HPLC column by a conventional reversed-phase process.

Another concurrent effect in the ion-pair mode is the adsorption of ion-pairing agent on the surface of reversed-phase packing material. This causes a decrease of the available hydrophobic surface and its transformation into an ion-exchange type surface.

Ion-pair HPLC mode is a superposition of two competitive processes: ion-exchange and reversed-phase. Component retention is strongly dependent on the type of ion-pairing agent, its concentration, and most of all, on the history of the used column. The virgin reversed-phase (RP) column does show the hydrophobic selectivity in the ion-pair mode. However, with time, the adsorbent surface can become covered with a dense layer of adsorbed surfactant. This may irreversibly transform the RP column into an ion-exchange one.

During the last ten years, it was realized that all organic ionizable compounds show some specific hydrophobic interactions with reversed-phase stationary phases [6–8]. These relatively weak interactions offer significant HPLC selectivity in the separation of even related compounds. pH is a primary tool for controlling this selectivity through the change of the analyte ionization state.

5.1.2 Analyte ionization

A simplistic rule for determining the retention in reversed-phase HPLC is that the more hydrophobic the component, the more the component is retained. By simply following this rule one can conclude that any organic ionizable component will have longer retention in its neutral form than its ionized form. Ionization is a pH-dependent process, so we can expect a significant effect of the mobile phase pH on the separation of complex organic mixtures containing basic or acidic components.

Ionization of the analyte could be expressed by one of the following equilibria

$AH \Leftrightarrow A^- + H^+$ for acidic components,

$B + H^+ \Leftrightarrow BH^+$ for basic components

Equilibrium constants are usually written in one of the following forms:

$$K_a = \frac{[A^-] \cdot [H^+]}{[AH]} \tag{1}$$

Equation 2 is derived by applying the definition of pH to Eq. 1.

$$pK_a = pH + \log\left(\frac{[AH]}{[A^-]}\right) \tag{2}$$

A similar expression could be written for bases.

As we mentioned before, the compound in its ionic form is more hydrophilic, so it not only tends to have less interaction with hydrophobic stationary phase, it also tends to be more solvated with water molecules. This solvation causes significant decrease of the retention of ionic components.

Since the pK_a is a characteristic constant of a specific analyte, from the above equation one can conclude that relative amounts of neutral and ionic forms of the analyte could easily be adjusted by varying the pH of the mobile phase. Moreover, if the pH is about two units away from the component pK_a more then 99 % of the analyte will be in either ionic or neutral form, depending upon the direction of the pH shift. HPLC analysis are best carried out in these two regions where the analytes are predominantly in one form.

5.1.3 Ionization and HPLC retention in reversed-phase HPLC

Let us now discuss the primary dependence of the retention of the ionizable analyte on the pH of the mobile phase. For basic components it will have the form shown in Fig. 5-1.

The dependence of the retention of basic components on the pH of the mobile phase could be subdivided in three regions (Fig. 5-1):

A Fully protonated analyte (cationic form), low retention. The analyte is in the most hydrophilic form. Its interactions with the hydrophobic stationary phase are suppressed.

B Partial protonation region. Coexistence of two analyte forms (protonated and deprotonated) in the mobile phase in equilibrium causing bad peak shape and non-

Figure 5-1. General dependence of the retention of basic analytes on the pH of the mobile phase. The inflection point of the curve corresponds to the component pK_a of the component.

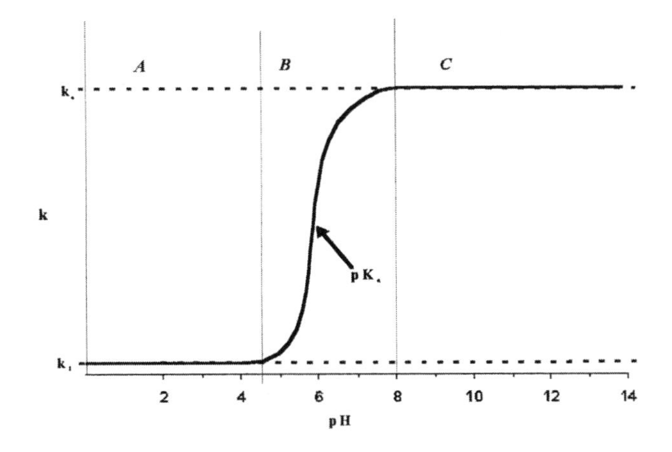

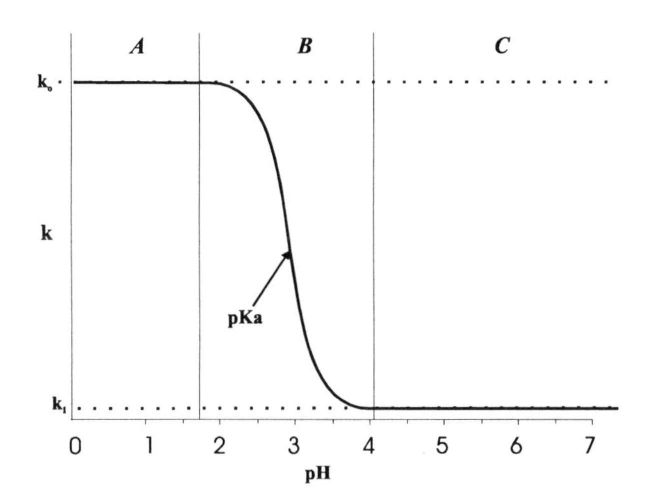

Figure 5-2. General retention dependence of acidic analytes on the pH of the mobile phase. The inflection point of the curve corresponds to the component pK_a.

stable retention. Since the analyte in the neutral form has much stronger retention its molecules tend to go onto the stationary phase more often and sit there longer. This causes a shift of the ionization equilibrium in the mobile phase towards a formation of deprotonated molecules and a further increase of overall retention of the front of the chromatographic band. The overall process depends on the superposition of the ionization and adsorption processes and their relative kinetics. Usually, a slight change of the pH of the mobile phase greatly shifts the retention.

C The analyte is in its neutral form (the most hydrophobic) and shows the longest retention.

The same type of retention curve may be obtained for acidic components, but their retention dependence will be the mirror image of that for basic analytes (Fig. 5-2).

These retention profiles seen in Figures 5-1 and 5-2 could be described by the following equations as a function of eluent pH and analyte pK_a [9].

$$k = \frac{k_0 + k_1 \frac{[H^+]}{K_{a(B^+)}}}{l + \frac{[H^+]}{K_{a(B^+)}}} \tag{3}$$

or

$$k = \frac{k_0 + k_1 \exp[2.3(pK_a - pH)]}{1 + \exp[2.3(pK_a - pH)]} \tag{4}$$

where k_l is the retention factor of the ionized form (protonated form for base and anionic form for acid, represented by the lower plateaus in Figs. 5-1 and 5-2), k_o is a retention factor of neutral forms for acids and bases (higher plateaus in Figs. 5-1 and 5-2); k is the current retention factor at a given pH; K_a is the analyte ionization constant.

Regions A for basic and C for acidic components show very low, if any, retention variation with a change of the pH of mobile phase. Methods employing a mobile phase pH which corresponds to these regions are generally more rugged.

On the other hand each region has its own drawback. One has to account for all possible effects when selecting the starting HPLC conditions and the direction of the separation development.

Effects in region A

- Basic analytes show relatively low retention (analyte in its ionic form).
- Acidic analytes show extremely long retention times (analyte in its neutral form).

Effects in region B

- Significant loss of apparent efficiency for both acidic and basic analytes. Peaks are broad and sometimes have weird shape, usually tailing or fronting.
- Very unstable retention. Minor changes in pH or composition of the mobile phase will significantly shift retention.
- Minor changes of the eluent composition can cause a significant change in selectivity.

Effects in region C

- Very long retention for basic analytes; this requires working with high organic concentration of the mobile phase where pH adjustment may not be efficient.
- Silica is soluble at high pH. If the column has some accessible silanols (which all columns usually do) high pH may cause steady degradation of the packing material. This brings a loss of the efficiency due to the formation of voids within a column, or steady change of component retention.
- If the mixture of analytes contains some acidic components, these components will be in anionic form at high pH. Organic analytes in their anionic form are usually strongly solvated and may be completely excluded from the pore space of the

packing material. This causes very early elution, usually not adjustable by the eluent composition.

5.1.4 Factors that should be considered prior to method development

Drawbacks that may be encountered when working at high pH.

At high pH, the stability of the packing is severely affected by the dissolution of the bonded phase [10]. Typically, reversed-phase packing materials that are currently available are stable up to about pH 8.5. Silica is soluble at high pH, and this may cause a steady dissolution of the column packing. Therefore, retention time reproducibility will be affected by the steady degradation of the stationary phase. For acidic and basic components there may be a decrease in retention on a column that has been exposed to high pH.

If it is necessary to work at high pH because compounds are known to degrade at low pH, what could be done to obtain a more rugged method?

It is still possible to work at these high pHs. Recently reversed-phase packings have been made more stable under high pH conditions. Bonding density of these packings is greater and the dissolution of the silica is less pronounced. Also, decreased operating temperatures (< 40 °C) are suggested when working at high pH. Therefore, a higher bonding density silica and lower temperatures would permit the chromatographer to work at high pH for the analysis of basic and acidic components if necessary.

In summary, HPLC analysis of basic analytes is more beneficial in a low pH region where these components are fully protonated and the problems associated with running with very high pHs in the mobile phase may be avoided. The elution of acidic and hydrophobic neutral components may be achieved by employing a gradient after basic components have been separated at low pH values.

5.2 Method development

5.2.1 General approach to method development

I know that I do not have any ionic or ionizable components in my mixture.

- Do not attempt to adjust your mobile phase pH. Your best choice will be to use an acetonitrile/water mixture. Changing the pH of the eluent will not effect the retention of unionizable components.
- Suitable retention factors may be obtained solely by modifying the organic composition and the employment of gradients.

My mixture contains weak acids and weak bases.

Weak acids are negatively charged when ionized, usually having a pK_a of 3 or higher. Weak bases are positively charged when ionized usually have a pK_b of 9 or lower. Remember, pK_a = 14 – pK_b. If your mixture contains weak acids or bases, you may add 0.5 ml of phosphoric acid to 1 l of water. This will bring the pH of your mobile phase down to ~2.3. If any ionizable components are within the mixture when the pH has been adjusting to below 3, in 99 % of the cases the analytes will be in the

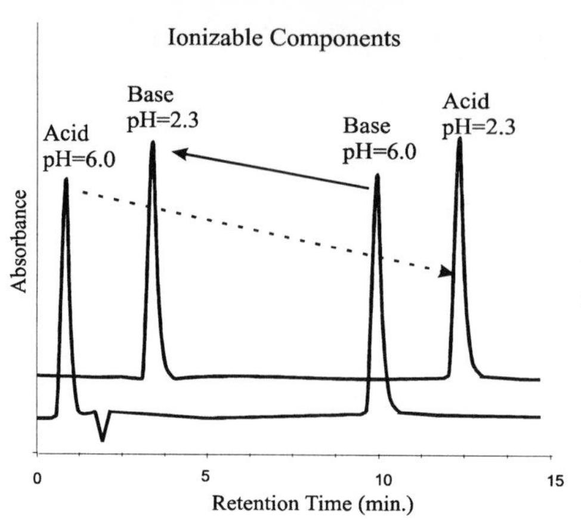

Figure 5-3. Effect of pH of the mobile phase on the retention of ionizable compounds.

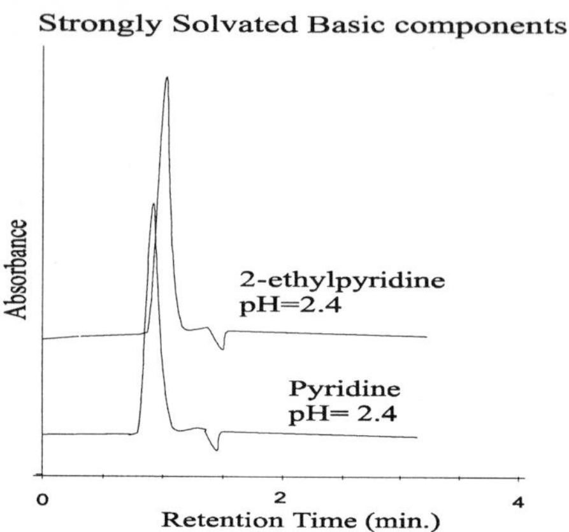

Figure 5-4. Effect on retention of strongly solvated small hydrophobic basic compounds. Column 15 × 0.46 cm Zorbax XDB-C18 with void volume of 1.4 ml; mobile phase: aqueous acetonitrile (80:20) adjusted with phosphoric acid, pH = 2.4, flow rate 1.0 ml/min, 25 °C, UV 254 nm, 1 µl sample injection.

fully protonated region (A) for bases and in the neutral region (A) for acids (refer to Figs. 5-1 and 5-2). Figure 5-3 illustrates the change in retention with change in pH.

What may happen if I analyze my weak basic components at low pH?

Typically, at low pH, the elution of the basic components is fast. This is advantageous to the chromatographer since the desired separation of basic components may be obtained in the shortest possible time. However, this retention may be too low and analytes will elute even at the void volume. The basic analytes in their cationic (protonated) form can be solvated, and therefore are hydrophilic. The hydrophobic nature of protonated basic analytes and their solvation will effect the elution. Therefore, if a basic analyte has a small molecular size and low hydrophobicity it may be strongly solvated and elute at the void volume.

Void volume is the total volume of the liquid phase inside the column. Highly solvated analytes form very hydrophilic molecular clusters which may not penetrate into the hydrophobic space inside the reversed-phase adsorbent particles. This will cause their fast elution primarily through the interparticle space with very low retention volume.

For example, small basic analytes such as pyridine and 2-ethylpyridine may elute before the void volume (Fig. 5-4). Other more hydrophobic analytes such as aniline and *N,N*-dimethylaniline under the same experimental conditions were eluted after the void.

The decrease of the organic content in the mobile phase, which usually significantly increases the analyte retention, does not have any effect on the elution of those excluded compounds. In these cases, some mobile phase additives which affect the analyte solvation may have to be employed to increase retention.

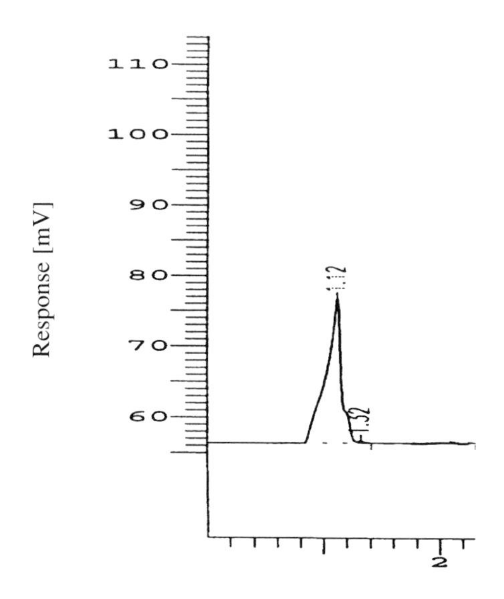

Figure 5-5. Retention of benzoic acid from 70/30 MeCN/H_2O, pH 6, on Prodigy ODS2 column with void volume of 1.7 ml.

What may happen if I analyze my weak acidic components at high pH?

Typically, at a high pH acidic components are ionized and highly solvated. Their retention is very low and often they may elute before the void volume. Figure 5-4 shows one practical example of what may happen with acids at high pH. Acidic analytes in their anionic form are even more strongly solvated than protonated bases. An ionic exclusion effect may be seen when your solvated analyte will not be able to penetrate inside the pore space of the packing material. This will cause its elution with very low exclusion volume as if it were a polymer. Figure 5-5 shows elution of benzoic acid at pH ~6.0 at 70/30 MeCN/Water on a Prodigy-ODS2 column (150 × 4.6 mm). The pK_a of benzoic acid is 4.2. The working pH is approximately 2 pH units greater than the pK_a. Therefore at this pH benzoic acid is fully ionized. Benzoic acid elutes in 1.12 min while an average void volume for this type of columns is 1.7 ml suggesting that it is excluded from the inside pore space of the packing material.

Another problem with using high pH is possible degradation of packing material. Silica is easily soluble at high pH and you can sometimes wash your packing material out of the column. One can argue that many manufacturers offer special "base deactivated" or other special columns stable at very high pH. We can make a simple analogy between a sports car for racing and a normal car. No one would buy a racing car for everyday travel; it is too expensive and after few hundred miles it needs significant repairs. On the other hand, a normal car cannot go at 200 mph, but can run steadily for years without much attention. Most reversed-phase HPLC columns are stable at low pH for many weeks if not years.

Are there any other drawbacks if I analyze my weak acidic components at high pH?

Bad peak shapes may be encountered. This may occur especially if the analyte is a small hydrophobic compound in which the hydrophobic part of the molecule is comparable to that of the acidic moiety. For larger hydrophobic acidic compounds this effect may not be predominant. The anionic form of the acidic compound is strongly solvated by water molecules. If the organic eluent modifier can participate in the analyte solvation, the solvation shell will have some hydrophobicity and may actually increase the retention. Figure 5-6a illustrates this effect with a small hydrophobic acidic compound, benzoic acid. Similar effects have been seen for salicylic acid. Methanol may hydrogen bond and solvate the acidic analytes. Therefore when methanol is used instead of acetonitrile at these higher pHs, bad peak shapes may be obtained. As can be seen in Fig. 5-6a, the benzoic acid peak front starts prior to the void and extends to 10 min. In order to improve peak shape the pH was decreased to 2.5 (see Figure 5-6b). Decrease of the pH suppresses the ionization of the acid. This increases the hydrophobicity of the acidic analyte and the overall retention. It also makes peaks more narrow and symmetrical.

Conclusion: Use low pH for analysis of acids.

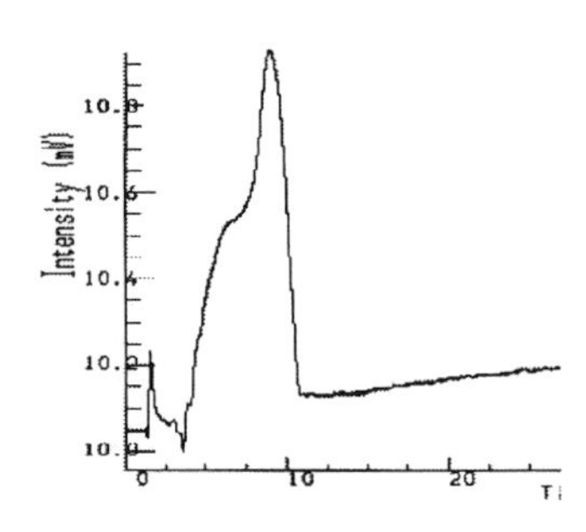

Benzoic acid in $MeOH/H_2O$
no buffer

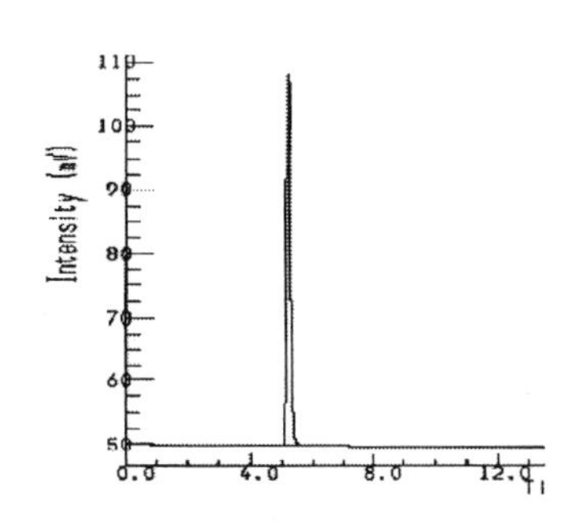

Benzoic acid in $MeOH/H_2O$
pH= 2.5

Figure 5-6. Effect of the organic eluent component on the solvation of acidic components and their peak shape. Column: Prodigy ODS2 column with void volume of 1.7 ml.

My components are weak acids and decreasing pH improves the peak shape but significantly increases the retention, up to 1 h for some analysis. How do I decrease the retention time?

To approach this problem the best way is to increase organic fraction in your eluent while keeping the same low pH.

5.2.2 Basic method development

I do not know the components of this mixture, but I have to develop a method as soon as possible. How do I start my method development?

One of the possible ways, to begin to develop a method, is to perform a sequence of experiments. Each step will help to determine information about certain components in your mixture. The general approach is to start with the simplest possible system and then add new parameters one at a time.

We will illustrate this approach on the model mixture of 6 unknown components as a sequence of steps, together with the explanation of the results after each change in the method.

Step 1. First try 80/20 MeCN/Water on a 15 cm reversed-phase column (select one of monomeric phase with the highest possible bonding density and that is endcapped).

Figure 5-7. Sample chromatogram of unknown mixture eluted with 80/20 MeCN/H_2O from reversed-phase column.

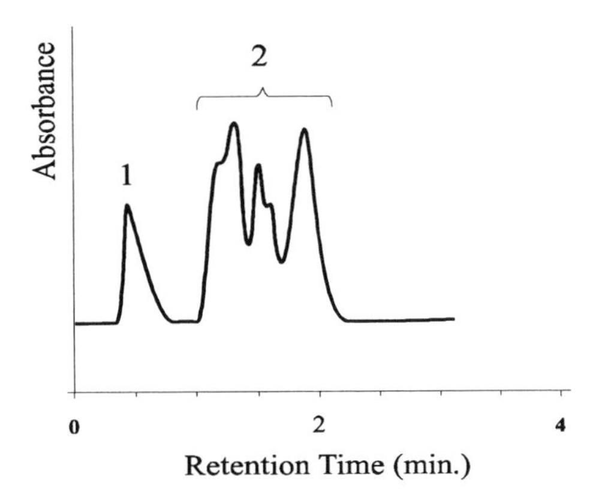

- High organic content will help you to ensure a complete elution of all your components. Do one run at 1 ml/min for 10–15 min.
- Column (150 × 4.6 mm) packed with 5 μm particles, usually has a ~1.7 ml void volume.
- If all your components are resolved then no further method development is needed.

Figure 5-7 shows the chromatogram of this unknown mixture.

- At peak *1* of Fig. 5-7, component 1 eluted very early (before the void) and may be either a polymer or acidic in nature.
- We do not yet know the nature of the components in region *2* of the chromatogram, but we could see that at least 6 components were present.

Step 2. Decrease the organic content. This will increase the retention of some components which have some hydrophobicity.

Figure 5-8 shows the change in the retention of our model mixture from the first run (80/20 MeCN/H_2O) to the second (50/50 MeCN/H_2O).

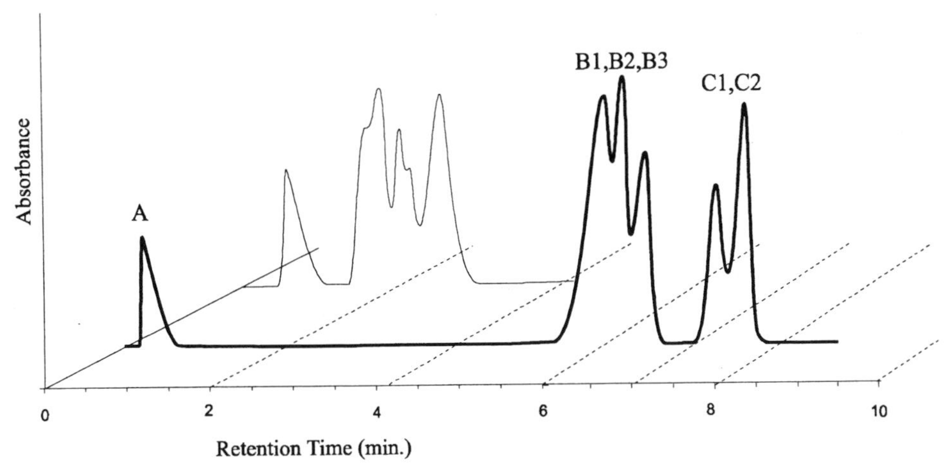

Figure 5-8. *Top* chromatogram: 80/20 MeCN/Water, pH 6.0; *bottom* chromatogram: 50/50 MeCN/Water, pH 6.0.

- Decreasing the organic content partially resolved some components in the second group of peaks.
- These components are denoted B1, B2, B3 and C1 and C2.
- Look at the peak shapes from the last run. Any broad or tailing peaks most probably indicate ionizable components.

Hints:

- Check the Photo diode array (PDA) peak purity of any suspicious peaks. Take spectra at five points across the peak. All the UV spectra should overlay if it is a pure peak. This may help you identify any coeluting components.
- Do not have PDA? Simply calculate the number theoretical plates for each peak. A significant drop in N (efficiency) for any peak compared with neighboring ones (by two or more times) indicates possible coelution.

Step 3. Now let us determine if ionizable components are present. For example, decrease the pH by adding 0.3 ml of phosphoric acid to 1 l HPLC grade water, pH ~4.0. Note, that you should actually measure the resulting pH even for preliminary experiments. Figure 5-9 illustrates the changes in the retention of a model mixture at pH 4.0.

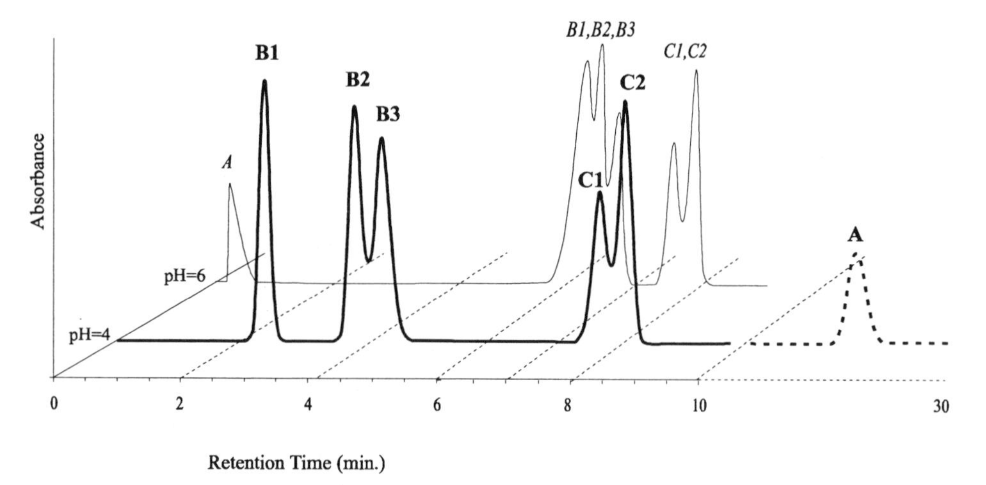

Figure 5-9. *Top* chromatogram: pH 6; *bottom* chromatogram: pH 4.

- It can be seen that peak A increased its retention. This is most likely an acidic compound. Decreasing the pH from 6 to 4 made the acidic analyte neutral and hence more hydrophobic. This confirms the nature of compound A.
- Some peaks can decrease their retention, sometimes also improving their peak shape (B1, B2, B3); these components are basic and become protonated and charged at a low pH.
- The retention of compounds C1 and C2 did not change. Compounds C1 and C2 may be nonionizable since they are not influenced by changes in the eluent pH.
- However, it is possible that analytes C1 and C2 may be acidic or basic since some components with very low or very high pK_a may not change their retention in the pH range 4–6.

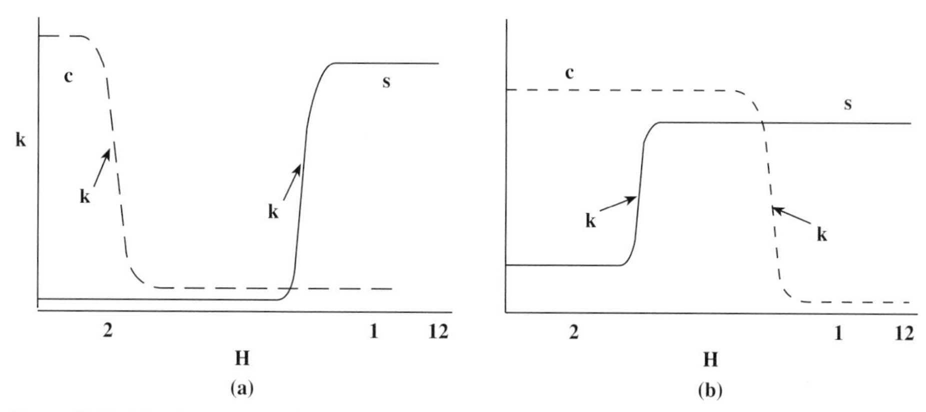

Figure 5-10. The dependence of retention on pH for acidic and basic compounds for which there was no effect between pH 4 and 6.

- The basic components may have high pK_a values and may be already protonated at pH 6, the approximate pH of MeCN/water mixture.
- The acidic components may have very low pK_a values and may already be in their anionic format pH 6.
- It is possible that the acidic or basic nature of components C1 and C2 were not manifested by lowering the pH from 6 to 4. For example, this would occur for an acidic component with a pK_a of 2 or lower or for a basic component with a pK_a of 8 or higher. Figure 5-10a shows theoretical curves of retention factor vs pH for these acidic and basic compounds. This could also occur for acidic compounds with a very high pK_a and bases with a very low pK_a. Figure 5-10b shows these theoretical curves. The retention factors for these theoretical acids and bases in Fig. 5-10a and b are very similar in the pH range 4–6. It can be seen in Fig. 5-9 that, when the pH was lowered from 6 to 4, this did not provide any specific information on the nature of the compounds.

Step 4. Let us check whether the compounds C2 and C1 whose retention did not change are acidic, basic or unionizable compounds in nature. We can do this by increasing the pH of the mobile phase. This is to be done solely for the purpose of determining the nature of C2 and C1. The pH can be adjusted to 7.5 by adding dilute NaOH to the mobile phase. Figure 5-11 shows the chromatogram of our mixture at pH 7.5.

- At this high pH, the acidic component (A) elutes before the void volume again as it did at pH 6, step 1. From the introductory discussion one may deduce that an increase of the pH of the mobile phase will ionize the acidic components and will result in a significant decrease in their retention. Even if no acidic components were identified in our mixture, it is not recommended to work at high pH because of silica solubility (see discussion above).
- The retention of compounds B1, B2 and B3 is similar at pH 6, indicating that these bases are in their neutral form.
- You can see in Fig. 5-11 that the C2 peak did not change its retention when the pH was adjusted from 6 to 7.5, indicating that it is most probably a neutral compound.

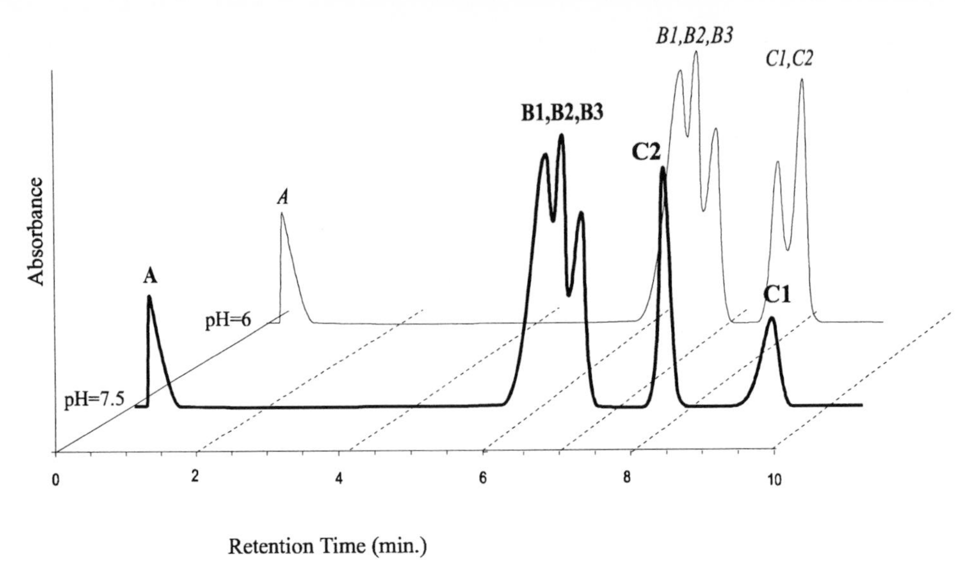

Figure 5-11. *Top* chromatogram: pH 6.0; *bottom* chromatogram: pH 7.5.

- However, the retention of compound C1 increased, indicating that it is a basic compound with a high pK_a. Also, some fronting is observed, which could be an indication that this pH is close to the pK_a of compound C1.

5.2.3 Method optimization

How can the conditions be optimized in order to obtain the desired resolution between components?

Points to consider

- A decrease in pH may be suitable in order to resolve the basic compounds B1, B2 and B3.
- After further lowering the pH it may be necessary to decrease the organic content at the initial conditions, since the basic compounds may otherwise elute very early.
- After ensuring adequate retention of the basic components at low pH, the acidic and nonionizable compounds maybe eluted using a gradient.
- In order to obtain a more rugged method, a buffer such as sodium dihydrogen phosphate should be employed. Minor variations in pH or temperature will not significantly influence the results in the presence of a buffered system. Phosphate buffers are useful since they offer a buffering capacity over a large pH range and are UV transparent from 200–400 nm.
- The aqueous portion of the mobile phase may be a sodium dihydrogen phosphate buffer adjusted with perchloric acid to pH 2.5. Perchloric acid has an advantage (to be discussed later) over other acids for this purpose.

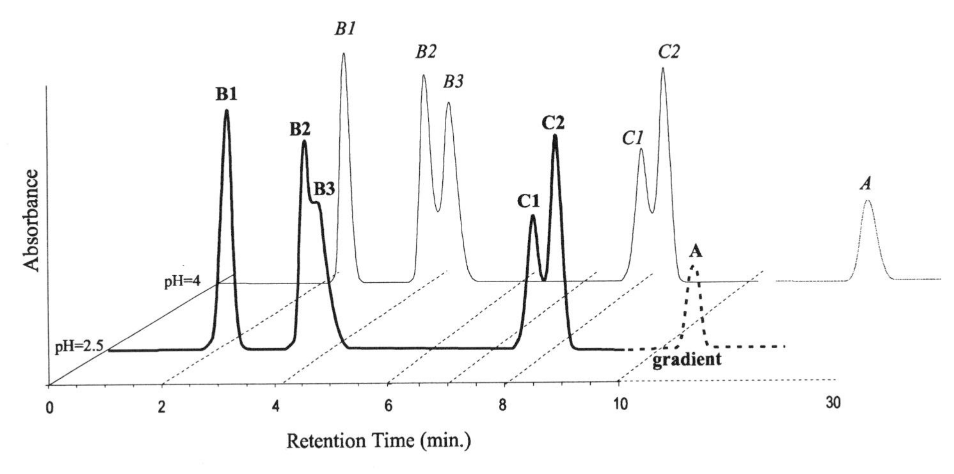

Figure 5-12. *Top* chromatogram: pH 4.0; *bottom* chromatogram: pH 2.5.

Step 5. Bearing these points in mind, let us see what the retention of the components will be at pH 2.5. Method conditions are 50:50 MeCN:sodium dihydrogen phosphate buffer adjusted with perchloric acid to pH 2.5 for 10 min, and then a linear gradient to 80:20 MeCN:water for 20 min. Total run time: 30 min; flow rate: 1.0 ml/min.

- The decrease in pH from 4.0 at step 3 to 2.5 (Fig. 5-12) further decreased the retention of the three basic components B1, B2 and B3. Their decrease in retention is dependent upon their ionization. B2 and B3 are starting to co-elute.
- The retention of compound C2 still did not change. It is unaffected by pH because it is most likely unionizable.
- The acidic component A at this low pH is neutral and therefore mainly unsolvated and hydrophobic. We were able to elute compound A with a capacity factor less than 10 by running simple linear gradient.
- If other more hydrophobic acidic or neutral compounds had been present in the mixture they may have been resolved because of the differences in their hydrophobic interaction with the phase.

Step 6. Further decrease the pH to see if better resolution between the basic compounds in obtained. The change in pH should be small since small pH changes could cause large changes in retention of the basic compounds. The pH of the mobile phase can then be adjusted to pH = 2 with dilute perchloric acid and we get the resulting chromatogram in Fig. 5-13.

- All peaks are resolved.
- The retention times of compound A (acidic) and the unionizable compound C2 remained the same at pH 2.5 and pH 2.0.
- There is a reversal in the order of elution of B2 and B3 and of C1 and C2.
- Peaks B1 and B3 decreased their retention with decrease in pH, illustrating that as the compounds became more ionized they became more polar and less retained. Since compound B2 is also basic we may expect a retention decrease as in the case of compounds B1 and B3. Also, we expect no change in retention of C1since it is

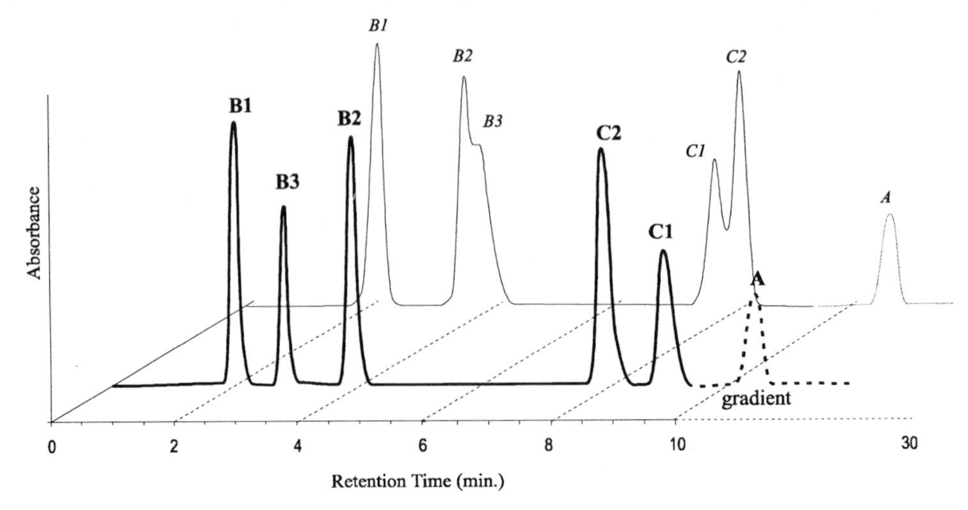

Figure 5-13. Top chromatogram: pH 2.5; *bottom* chromatogram: pH 2.0.

in its fully ionized state. However, the retention of components B2 and C1 increased upon further decrease of the pH from 2.5 down to 2.

What is the cause of this increase in retention of the basic compounds B2 and C1 upon lowering the pH?

At pH 2, compound C1 and compound B2 are fully protonated and the perchloric acid is in its fully ionized (anionic) form. The increase in retention in the pH range from 2 to 2.5 is governed by ion association, between protonated basic analyte and negatively charged perchlorate ion. This ion association causes a desolvation of water molecules around the protonated basic analyte. Once the protonated basic analyte is less solvated by water it has a more hydrophobic nature, and for this reason the retention on the reversed-phase packing is increased.

Why did this not occur for compounds B1 and B3 at pH 2 even though they are basic?

As was seen in Fig. 5-12 (pH 4 and pH 2.5) and Fig. 5-13 (pH 2.5 and pH 2.0), the retention of the basic compounds B1 and B3 decreased, indicating at these pHs the compounds were not fully protonated. In order for this ion association to cause a noticeable increase in the retention of the basic analytes, the basic compounds must be fully protonated. The basic analytes are solvated by the water and solvent molecules to a greater extent when fully protonated than when in their neutral forms. Partially ionized basic analytes contain some of their neutral form as well as their protonated form, and this ion association effect will not be the predominating factor in determining the retention. The decrease in retention for compounds B1 and B3 was governed by ionization. As the pH decreased, the compounds were becoming more ionized, more polar, and were less retained. The increase in the pH range 2.5–2 for compounds B2 and C2 is governed by ion association.

5.2.4 Conclusion

In Steps 5 and 6 of the method development we showed how one can determine the nature of the analytes in an unknown mixture and also perform a preliminary optimization of an HPLC separation of the mixture. The approach described works in approximately 80 % of the cases. The other 20 % may require other types of optimization. This approach is for regular reversed-phase columns (C18 type), preferably with the highest possible bonding density. Other columns may introduce some specific interactions for certain analytes and may possibly simplify the development process for particular mixtures, but such a result should be regarded as the exeption rather than the rule.

Often methods developed using the approach described may require some fine tuning to increase the resolution of critical components or to make the separation method more rugged. Some specific steps in this fine tuning process, related to the mobile phase pH adjustment, will be discussed in the following Sections.

5.3 Method fine tuning

Do other acidic modifiers used to adjust the pH of the mobile phase affect the retention of these protonated basic analytes?

Yes, other acidic modifiers have shown the same effect. It was found that the highest effect on the retention of fully protonated analytes was obtained by the employment of perchloric acid, use of trifluoroacetic acid yielded results to a lesser extent and in some cases, phosphoric acid also had shown some effects. Also, other acids such as nitric acid, formic acid, acetic acid, propanoic acid, may also show similar effects.

As was discussed before, the counteranion of the acid actually effects the solvation of the analyte. Therefore certain properties of the counteranion may influence this change in solvation of the protonated basic analyte to a greater extent.

What are the properties of the acid that may affect the solvation of the basic analytes?

There are two main properties:

1. The hydrogen bonding ability of the acids (solvation).
2. The pK_a of the acid and its ionization at a particular pH.

5.3.1 Solvation of the acids

The less the ability of the acid to hydrogen bond, the less is its solvation in water. Hence, if the counteranion of the acid is itself less solvated, it may undergo ion association more easily and affect the solvation of the protonated basic analyte to a greater extent. The following are some acids typically used in HPLC and their solvation properties.

- *Phosphoric Acid*
  The phosphate counteranion is known to exhibit strong hydrogen bonding properties. It may act as a hydrogen donor through its hydrogen atoms and as an

acceptor through the phosphone group. Therefore, it is a highly solvated counteranion.

- *Trifluoroacetic acid*
  The TFA counteranion exhibits weak hydrogen bonding with water molecules. It acts as a proton acceptor through its carboxylate group and as a donor (weaker) through its fluorine atoms.
- *Perchloric acid*
  This is a strong acid that fully dissociates in water. Hydrogen bonding is highly unlikely. It is a less solvated counteranion than phosphate or trifluoroacetate.

5.3.2 Ionization of the acids

The optimal counteranion of the acid will have its negative charge dispersed or delocalized throughout its structure, thus having less ability to hydrogen bond. The lower the pK_a of the acid the stronger it is, and it may be ionized at pHs usually used for reversed-phase HPLC. However, weak acids may not be fully ionized at a certain pH of the mobile phase, and this will have an effect on their interaction with the basic analytes. Only the ionized form of the acid participates in the ion association. The following are some acids typically used in HPLC and their ionization properties.

- *Phosphoric Acid*
  The lowest pK_a of phosphoric acid is 2.1 and therefore at a pH of 2.1 it is only 50 % ionized. The negative charge is dispersed over two oxygen atoms. Further, decreasing the pH reduces the ionization of the phosphoric acid.
- *Trifluoroacetic Acid*
  The trifluoroacetate anion has a negative charge dispersed over two oxygen atoms and, furthermore, the dispersal of charge is even greater due to an electron-withdrawing effect of the fluorine atoms. TFA has a pK_a = 0.5 and is fully ionized at pHs above 2.5. Generally, it is fully ionized in the pH ranges typically used in reversed-phase HPLC.
- *Perchloric Acid*
  Perchloric acid has a very low pK_a and is fully ionized down to a pH of 1. The perchlorate counteranion has a single negative charge delocalized over all four of its oxygen atoms. It is the most polarizable of all the three counteranions discussed and is completely ionized throughout the whole pH range between 1 and 14.

Figure 5-14 shows the ionized forms of the acids and their expected solvation in water.

When these counteranions are present in the eluent and interact with the protonated basic analyte they tend to disrupt the solvation of the basic analyte. These anions can be classified as "chaotropic counteranions."

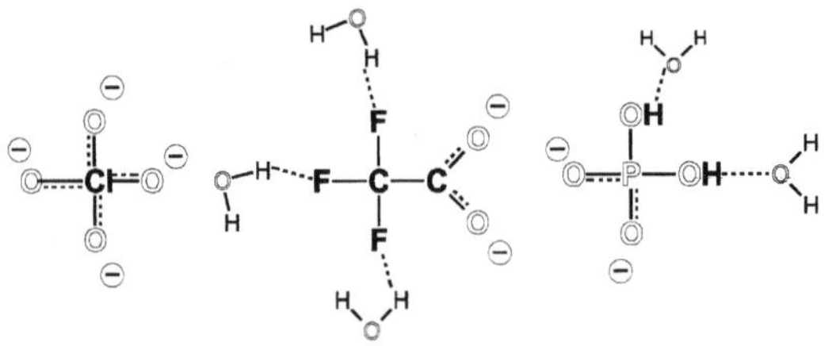

Figure 5-14. Ionized forms of three common acids used in HPLC. Note that these are only proposed hydrogen bonding interactions and are shown for illustration purposes.

5.3.3 Chaotropic effects

What are „chaotropic" counteranions?

Chaotropic anions are ions that have low localized charge, high polarizability and a low degree of solvation [11]. These ions change the structure (hydrogen bonding) of water in the direction of greater disorder [12]. The hydrogen bonding of water is broken down to varying degrees around the basic analyte because of its interaction with the chaotropic anion, which increases the hydrophobic nature of the basic analyte. Therefore, after subsequent ion association, these chaotropic counteranions will effect the retention of the fully protonated basic analytes. This phenomenological description of the increase of the retention of protonated basic analytes may be termed the *chaotropic approach*. The effect increases with addition of acid, which actually results in a decrease of pH and an increase of the concentration of counteranion of the acid in the mobile phase.

How does the use of different modifiers affect the retention of a basic analyte?

Figure 5-15 shows how the retention of 4-ethylpyridine varies with pH. The greatest increase in retention occurred when trifluoroacetic acid and perchloric acid were employed. The phosphoric modifier did not increase the retention of the 4-ethylpyridine significantly at decreasing pH values. Therefore the trifluoroacetate and perchlorate had a greater influence than phosphate as chaotropic counteranions at any particular pH. It was shown that this chaotropic effect began to occur at pHs less than 3. At all pHs greater than 4 the retention factors of the basic compound were similar and were independent of what type of modifier was used in the mobile phase.

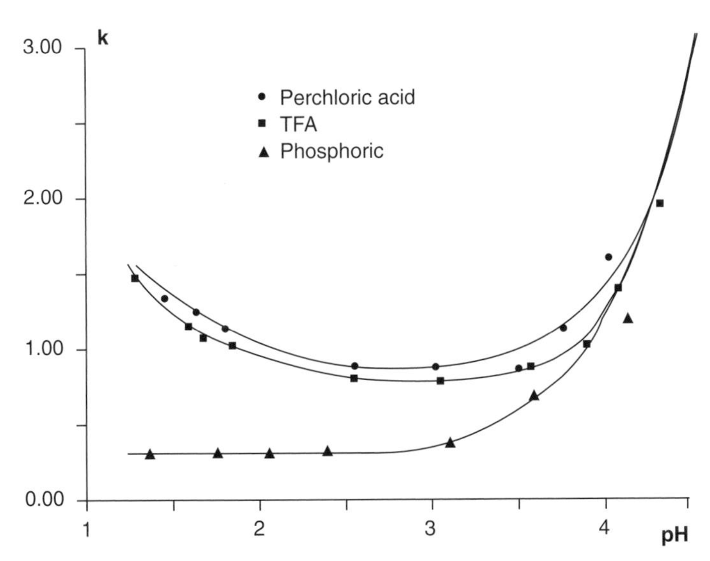

Figure 5-15. Comparison of retention factors for 4-ethylpyridine as a function of pH using three different acidic modifiers. Column 150 × 4.6 mm Zorbax XDB-C18. Mobile phase: acetonitrile - 10 mM sodium phosphate buffer adjusted with trifluoroacetic acid, (10:90); flow rate: 1.0 ml/min; 25 °C, UV 254 nm, sample: 1 µl injection.

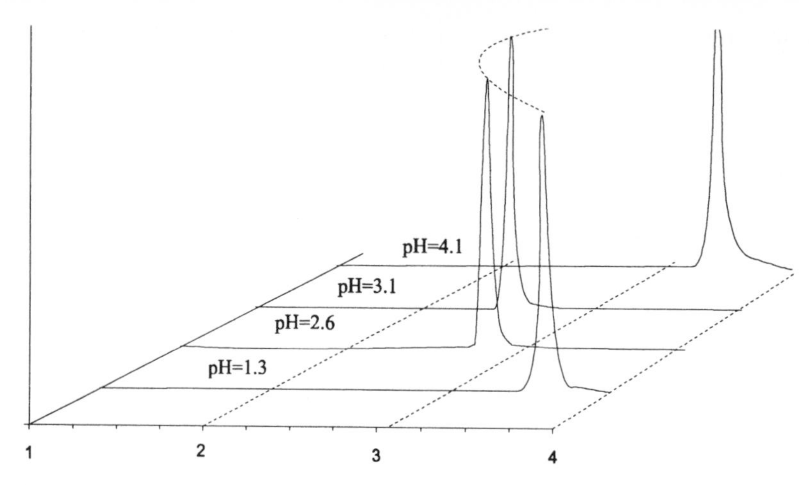

Figure 5-16. Chaotropic effect on retention of 4-ethylpyridine when TFA was used as the acidic modifier. Column 15 × 0.46 cm Zorbax XDB-C18; mobile phase: acetonitrile10 mM sodium phosphate buffer adjusted with trifluoroacetic acid (90:10); flow rate 1.0 ml/min; 25 °C; UV 254 nm; sample:1 µl injection. Retention at pH 4.1, 3.5; at pH 3.1, 2.6; at pH 2.6, 2.7 and at pH 1.3, 3.6.

In a low-retention pH range even a small change of pH will have a significant effect on the resolution since chromatographic peaks are narrow (see Fig. 5-16 overlay of 4-ethylpyridine retention).

At pH values below 3, 4-ethylpyridine did not show the theoretically expected plateaus of its retention dependence vs pH (Fig. 5-1). When pH adjustment of the mobile phase was achieved by addition of perchloric or TFA acids, 4-ethylpyridine increased its retention with increase of the amount of acid added.

Is the increase in retention for a protonated basic compound due to a decrease in pH or an increase in acidic modifier counteranion concentration?

It is an effect caused primarily by the concentration and type of counteranion. As a result of addition of acid there is an increase in the concentration of the counteranion of the acid and a simultaneous decrease in the pH. pH is a factor affecting the protonation of the analyte. Only when the analyte is protonated it can undergo ionic association with the counteranion of the acid that was used. Hence, when the protonated base interacts with the counteranion this leads to changes in its solvation and increase in its hydrophobicity. At higher concentrations of the acidic counteranion, the protonated basic analyte is desolvated to a greater extent. This ultimately causes an increase in analyte retention.

There are two ways in which the concentration of the counteranion may be increased:

1. Increase the amount of acid added
2. Increase the amount of added salt that contains the same counteranion as the acid.

In the first approach not only does the concentration change but so also does the pH. Therefore, with an increase in concentration of acid there is a decrease in pH. This can

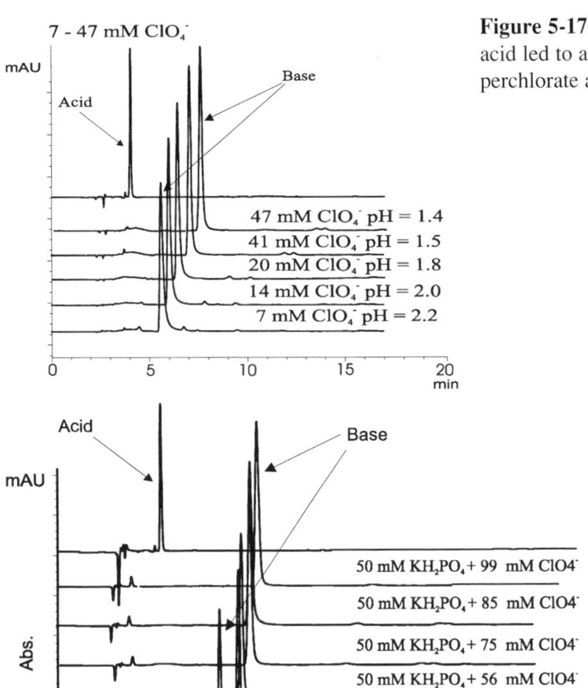

Figure 5-17. Increased addition of perchloric acid led to an increased concentration of perchlorate anion and a decrease in pH.

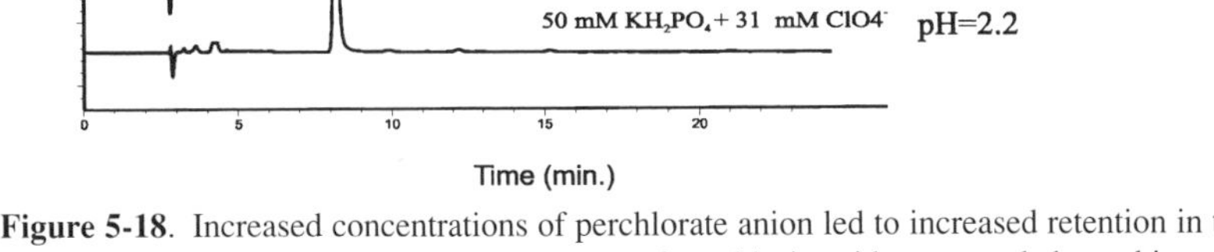

Figure 5-18. Increased concentrations of perchlorate anion led to increased retention in the presence of a phosphate buffer system. Increased amounts of perchloric acid were needed to achieve the same pH values as in Fig. 5-17 because of the buffering capacity of the phosphate buffer.

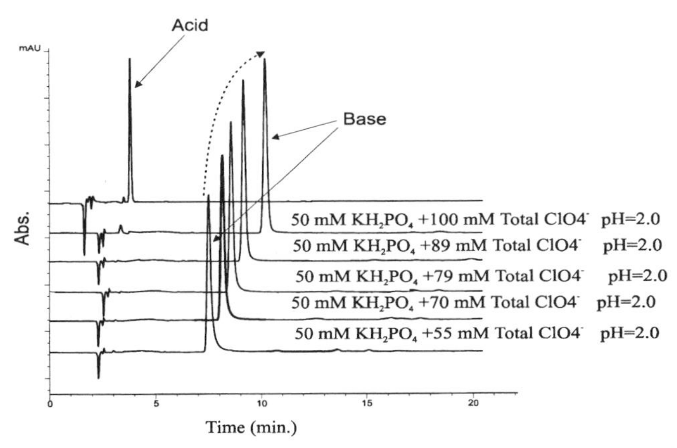

Figure 5-19. Increase in retention is observed with increased concentration of perchlorate at constant pH. Starting concentration of perchlorate anion from perchloric acid is 14 mM in order to reach pH 2.0. Further addition of sodium perchlorate salt led to increased retention of the basic analyte.

be seen in Figs. 5-17 and 5-18. In Fig. 5-17, the concentration of the perchlorate anion was increased by addition of perchloric acid. In Fig. 5-18, a phosphate buffer was employed. Therefore, because of the buffering capacity, higher concentrations of perchloric acid were needed to obtain the same pHs as in Fig. 5-17, and as a result greater retention times were obtained. On the other hand if the second approach is used the addition of the salt will increase the concentration of the counteranion at a constant pH (Fig. 5-19). As can be seen from Fig. 5-19 the increase in the perchlorate anion concentration by simple addition of $NaClO_4$ caused the most significant increase of the analyte retention without a change in pH. This ultimately shows that the increase in retention is independent of pH and is solely dependent on the concentration of the perchlorate anion.

The following is a graph showing the retention factors as a function of concentration of perchlorate anion obtained in Figs. 5-17 to 5-19 under the three different experimental conditions. It can be seen that, regardless of pH, the counteranion concentration is the determining factor that affects the solvation and ultimately the retention of the analyte.

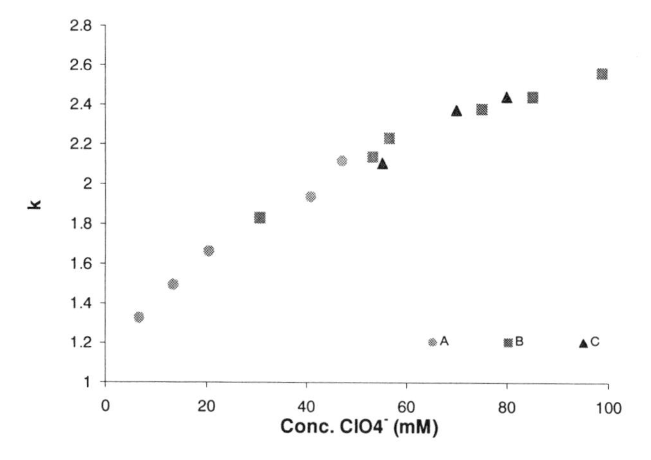

Figure 5-20. Retention factor as a function of concentration of perchlorate anion on basic analyte. A: pH of mobile phase adjusted with perchloric acid, B: 50 mM sodium phosphate buffer with pH adjustment using perchloric acid, C: pH of mobile phase adjusted with perchloric acid and $NaClO_4$ salt then added.

How different is the chaotropic effect for different analytes?

Figure 5-21 illustrates the chaotropic effect for several basic analytes. The effect of the chaotropic agent on the disruption of the basic analytés solvation shell is dependent on the type and position of substituents. At the various concentrations, the effect on the retention was different for the analytes of different stereochemistry and led to increased resolution of certain components of similar structure.

These were all done with perchloric acid as the modifier. It is considered to be a strong chaotropic agent. Weak chaotropic counteranions will produce the same type of retention dependence, but the overall effect of weak chaotropes on the analyte retention is much less pronounced.

At a certain pH when different acidic modifiers are employed the counteranion concentrations are not the same. How can one compare the effect of the chaotropic counteranion on the retention of basic analytes?

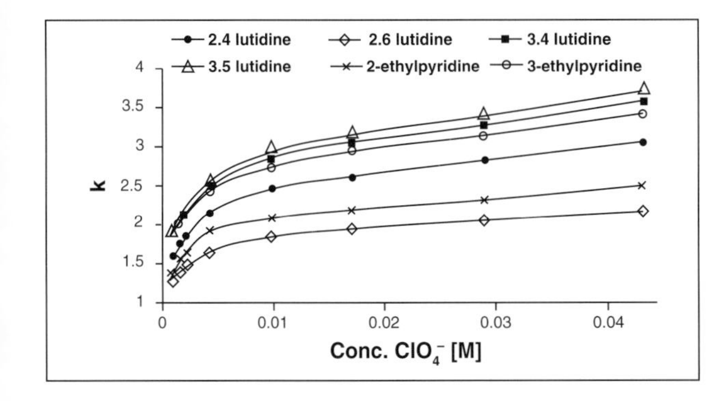

Figure 5-21. Change of the retention of basic analytes at low pH with increase of the concentration of counteranion. Concentration region 0.08 mM to 44 mM perchlorate anion. Column: 15 × 0.46 cm Zorbax XDB-C18; mobile phase: methanol:aqueous adjusted with perchloric acid pH 1.4–2.9 (90:10); flow rate: 1.0 ml/min.; 25 °C; UV 254 nm; 1μl injection.

It is extremely important that the retention factor is plotted against the concentration of the acidic counteranion employed at particular pHs. If a weak acid is employed, it will be ionized to different degrees at different pHs. The effects of the concentration on the retention factor may actually be hidden if the ionization of the acid is not taken into consideration. Therefore, the proper way to investigate the effect of the strength of the anionic chaotrope upon the analyte retention is to compare at the same concentration of counteranion and not pH. Figure 5-22a shows the retention factor of a basic compound, 2-propylpyridine, plotted against pH when two different acids were used to adjust the pH of the mobile phase. Figure 5-22b shows the retention of the same analyte plotted against the concentration of trifluoroacetate and perchlorate.

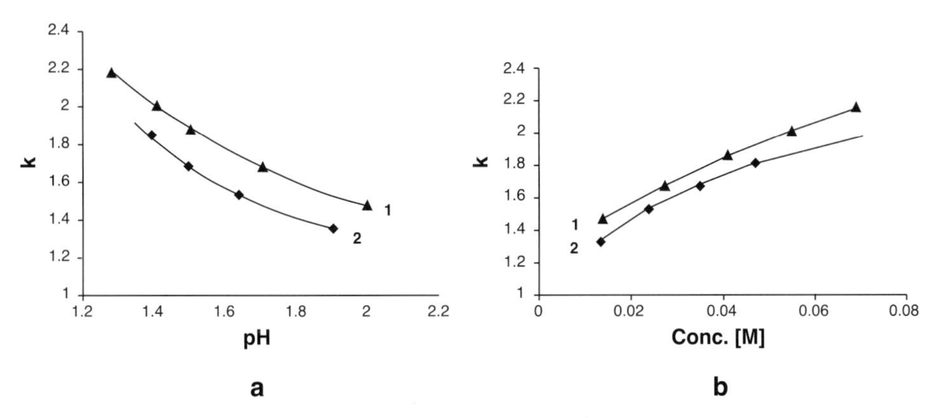

Figure 5-22. Comparison of the chaotropic effect on 2-propylpyridine caused by perchlorate and trifluoroacetate counteranions plotted against mobile phase pH (a) and counteranion concentration (b). Line 1 is for perchloric acid and line 2 is for trifluoroacetic acid. Column 150 × 4.6 mm Zorbax XDB-C18; mobile phase: acetonitrile-HPC grade water adjusted with trifluoroacetic acid or perchloric acid, pH 1.3–1.9 (10:90); flow rate: 1.0 ml/min; 25 °C, UV 254 nm; sample: 1 μl injection.

When the retention factor is plotted against the pH (Fig. 5-22a), the increase in retention caused by the perchlorate modifier seems to be more significant. However, the graphical representation of retention factor vs concentration (Fig. 5-22b) actually defines the strength of the chaotropic anion. This observed increase in retention shows that, for similar concentrations of the counteranion of the acid, perchlorate is more chaotropic than trifluoroacetate. It disrupts the solvation of the basic analyte to a greater extent than the trifluoroacetate counteranion and hence increased retention factors are obtained. Therefore, the concentration of the counteranions must be known in order to properly compare the retention factors obtained when different types of counteranions are employed in the mobile phase.

How is the concentration of the counteranion of a strong acid calculated?

If a strong acid is used, it is assumed that it fully dissociates in the aqueous portion of the mobile phase. In the case of perchloric acid there will be no $HClO_4$ molecules but only H^+ and ClO_4^- ions present. Usually the K_a (dissociation constant) values of strong acids are >1. Other acids in this class include hydrochloric, hydrobromic, nitric, and sulfuric.

The following is a simple general explanation for the calculation of the concentration of the counteranion for your convenience. So instead of going back to those old and misplaced general chemistry textbooks you can use the following simple guide in your everyday work. The number of moles of perchlorate would be calculated directly from the number of ml of acid added, not neglecting the fact that the perchloric acid usually is not in its pure form. Usually, it is 65 – 70 % by volume in water. The density of 70 % v/v perchloric acid in water is 1.664 g/ml. The density of the pure perchloric acid may be calculated directly knowing the density of water to be 1.0 g/ml:

$$d_{\text{acid}} = \frac{d_{70\%\,\text{acid}} - 0.3 d_{\text{H}_2\text{O}}}{0,7} = 1.95\,\text{g/ml}$$

The resulting density of perchloric acid is 1.95 g/ml. Knowing the volume of the aqueous portion to which the acid is added, the concentration in mol/l may be calculated. The following equation may be used to calculate the concentration for a strong acid:

$$\frac{V_{\text{acid}} \cdot C_{\text{acid}} \cdot d_{\text{acid}}}{\text{MW} \cdot V_{\text{aq}}} = C$$

where V_{acid} is the volume of acid added, C_{acid} is the concentration of acid expressed in terms of volume % of the acid, d_{acid} is the density of the pure acid, MW is molecular weight of the strong acid, and V_{aq} is the volume to which the acid was added; C is the concentration of perchloric acid in mol/l.

The following is an actual example of how to calculate the concentration of perchlorate anion. 1.0 ml of perchloric acid (70 % v/v in water) was added to 2.0 l of HPLC grade water.

V_{acid} = 1.0 ml, d_{acid} = 1.95 g/ml, MW = 100.5 g/mol, V_{aq} = 2.0 l, C_{acid} = 0.70

$$\frac{1.0 \cdot 0.70 \cdot 1.95}{100.5 \cdot 2.0} = 0.0068\ \text{M or } 6.8\ \text{mM perchlorid acid.}$$

In Table 5-1, the perchlorate anion concentration and the corresponding pH value are given for a certain volume of perchloric acid added to 1 l of HPLC grade water.

Table 5-1. Perchlorate ion concentration and pH at various concentrations of perchloric acid.

ml. $HClO_4$ added	Conc. $[ClO_4^-]$ mM	pH
1.0	13.6	1.99
2.0	27.2	1.70
3.0	40.8	1.50
4.0	54.3	1.39

How is the concentration of counteranion from a weak acid calculated?

This situation is more complicated than that for a strong acid. Since a weaker acid is not completely dissociated, an equilibrium calculation must be performed to find $[H^+]$ and $[A^-]$. The concentration of a weak acid must be known, especially if the working pH is within a range where the acid is only partially ionized. This ionization must be taken into consideration in the calculation. Figure 5-23 shows the ionization of perchloric acid and trifluoroacetic acid at various pHs. Trifluoroacetic acid is a weaker acid than perchloric acid.

- Note, that perchlorate counteranions were 100 % ionized since it is a strong acid and fully dissociates in water.
- A decrease from full ionization of trifluoroacetate is obtained when the pH is below 2.5.

The molar concentration of the ionized counteranion of the acidic modifier must be taken into consideration since different molar amounts of acid must be added to obtain the same pH with the different acids. Trifluoroacetic acid does not fully dissociate at low pH. Not only should the strength of the acid be considered but also, whether it is monoprotic or polyprotic. For the first case we will consider a monoprotic weak acid such as trifluoroacetic.

The concentration of the counteranion may be calculated from the K_a and density of the acid, together with the actual volume of the acid added to a given volume of HPLC

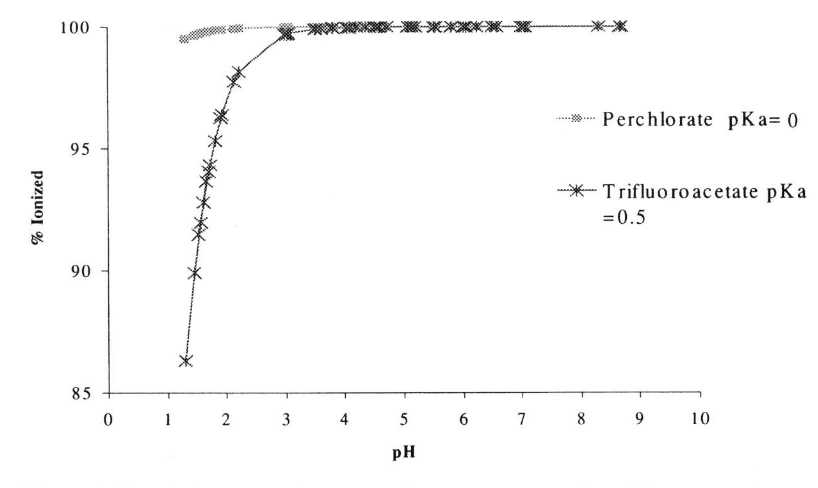

Figure 5-23. The ionization of a strong acid and a weaker acid at different pH values.

grade water. The following is the procedure to calculate the concentration of trifluoroacetate.

The problem will be solved by using a series of steps.

1. The major species should be listed: CF_3COOH and H_2O.
   The trifluoroacetic acid will be present in its associated form and as its ions, since it is a weaker acid than perchloric acid.
2. The next step is to determine which of the major species in the solution has a dominant contribution of $[H^+]$. Remember that H_2O and CF_3COOH are both acids. Let us write the equations for the reactions in which H^+ is formed and their dissociation constants, K_a. The K_a may be calculated directly from the pK_a.

   $$pK_a = \log K_a \quad \text{The } pK_a \text{ of TFA is 0.5.}$$

   $$CF_3COOH \Leftrightarrow H^+ + CF_3COO^- \qquad K_a = 0.3$$

   $$H_2O \Leftrightarrow H^+ + OH^- \qquad K_a = 1.0 \times 10^{-14}$$

   Trifluoracetic acid is a weak acid, but it is still stronger than H_2O, and therefore the dissociation of CF_3COOH will make the predominant contribution to $[H^+]$.
3. The expression for the equilibrium of trifluoroacetic acid is:

   $$CF_3COOH \Leftrightarrow H^+ + CF_3COO^-$$

   Therefore $K_a = \dfrac{[H^+][CF_3COO^-]}{[CF_3COOH]}$
4. First we have to calculate the original concentration of trifluoroacetic acid:

   $$\frac{V_{acid} \cdot d_{acid}}{MW \cdot V_{aq}} = C$$

   V_{acid} is the volume of acid added, d_{acid} is the density of the acid, MW is the molecular weight of trifluoroacetic acid, V_{aq} is the volume to which the acid was added, and C is the resultant concentration in mol/l of acid in the aqueous phase. The purity of the acid should always be taken into consideration if it is not in its pure form. For example, let us assume that 0.7 ml of TFA was added to 1 l of water.
   V_{acid} = 0.7 ml, d_{acid} = 1.48 g/ml, MW = 114 g/mol, V_{aq} = 1 l
   Therefore our concentration is 0.0091 M.
5. We have to consider initial and equilibrium conditions.
   Initial conditions are concentrations of the species before any acid dissociation actually occurs.
   [HA] = 0.0091 M, $[A^-]$ = 0 M and $[H^+] \simeq 0$ M
6. Next, consider the change, denoted as x, required to reach equilibrium.

	$CF_3COOH \Leftrightarrow$	$H^+ +$	CF_3COO^-
Initial	0.0091	0	0
Equilibrium	0.0091-x	x	x

7. Substitute the concentrations at equilibrium conditions into the equilibrium expression and we obtain:

   $$\frac{x^2}{0.0091 - x} = K_a \rightarrow \frac{x^2}{0.0091 - x} = 0.3$$

This is a quadratic equation and its solution for x will be 0.0088 M or 8.8 mM.

$$x^2 + (0.3 \cdot x) - (0.0091 \cdot 0.3),$$

Therefore x is a concentration of H^+ as well as A^- which is CF_3COO^- in our case. Using the quadratic equation, more accurate results may be obtained for the concentration of the counteranion of the weak acid. Our concentration of x is 8.8 mM. This is a weaker acid than perchloric acid since it does not fully dissociate and the amount of trifluoracetate anions present will be dependent on the weaker acid's ionization at a certain pH.

Table 5-2 shows the concentration of trifluoroacetate corrected for its ionization and its corresponding pH after a certain volume of TFA was added to 1 l of water.

Table 5-2. TFA ion concentration and pH at various concentrations of TFA.

mL TFA added	Conc. CF_3COOH [mM] added	Conc. CF_3COO^- [mM] corrected for ionization	pH
0.2	2.6	2.6	2.6
0.7	9.1	8.8	2.1
1.0	13.0	12.5	1.9
2.0	25.9	24.1	1.6
3.0	38.9	35.0	1.5
4.0	51.8	45.3	1.4
5.0	64.8	55.1	1.3
6.0	77.8	64.6	1.2

The preceding example showed how to calculate the concentration of a weak monoprotic acid. The following describes the procedure for a polyprotic acid in a low pH region.

How do you calculate the concentration of a polyprotic acid such as phosphoric acid?

Since, phosphoric acid is a weak polyprotic acid that has three dissociation constants, four species (PO_4^{3-}, HPO_4^{2-}, $H_2PO_4^-$, and H_3PO_4) will coexist in equilibrium with one another although the concentration of some may be negligible at a particular pH. Therefore the dominating species at a particular pH will differ at different pHs. Figure 5-24 shows, the % of the ionized forms of phosphoric acid plotted against the pH.

Therefore in order to calculate the concentration of the dominant counteranion at a particular pH the following steps must be considered.

1. With the change of pH the relative concentrations of the different counteranions will change. If we are more than 2 units away from a particular pK_a of this polyprotic acid then the contribution of H^+ from that dissociation becomes negligible.
2. For example, at pH 2.0 the majority of the species of phosphoric acid exist as two forms: H_2PO_4 (44 %) and H_3PO_4 (66 %). The contributions from the hydrogen phosphate and phosphate are much less than 0.01 %. Therefore in this case only two species are present, H_3PO_4 and $H_2PO_4^-$, and only one equilibrium may be considered. At pH 2.0 the dissociation of H_3PO_4 will make the predominant contribution to $[H^+]$. At higher pHs the problem becomes more complicated and is

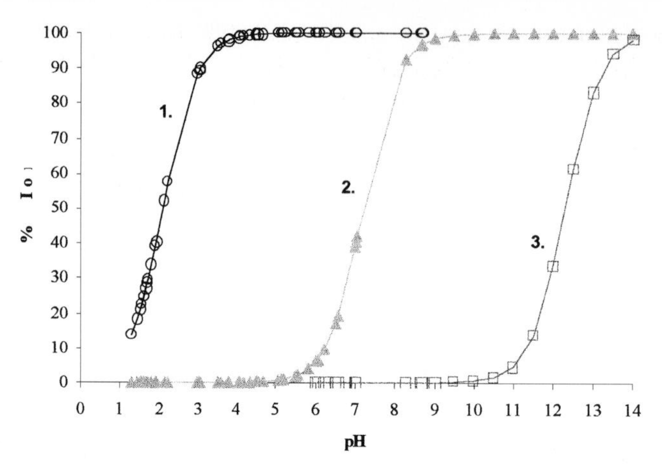

Figure 5-24. Ionization of a polyprotic acid H_3PO_4. 1. $H_2PO_4^-$, 2. HPO_4^{-2}, 3. PO_4^{-3}.

out of the scope of this discussion since we are focusing our method development in a low-pH range. The following equilibrium expression may be written:

$$K_a = \frac{[H^+][H_2PO_4^-]}{[H_3PO_4]}$$

The pK_a of phosphoric acid is 2.1 and the corresponding K_a is 7.94E-03.

3. The concentration of phosphoric acid is usually reported as a weight %, so the actual concentration will be:

$$\frac{V_{acid} \cdot C_{acid} \cdot d_{acid}}{MW \cdot V_{aq}} = C$$

where V_{acid} is volume of the acid added, C_{acid} is concentration of the acid expressed as the weight % of the acid, d_{acid} is the density of the 85 wt% phosphoric acid in solution, MW is molecular weight of phosphoric acid, V_{aq} is the volume to which the phosphoric acid was added, and C is the concentration of phosphoric acid. For example if we add 1.5 ml of phosphoric acid (85 wt% in water) to 1 l, V_{acid}=1.5 ml, C_{acid} = 0.85, d = 1.685 g/ml, MW = 98 g/mol, V_{aq} =1 l,

$$\frac{1.5 \cdot 0.85 \cdot 1.685}{98 \cdot 1} = 0.0219 \text{ M or } 21.9 \text{ mM of } H_3PO_4\text{ , pH} = 2.0.$$

4. Now the same steps, 5–7, that were carried out for trifluoroacetic acid in the previous section can be performed to calculate the concentration of dihydrogen phosphate counteranions where x is H^+, as well as A^- which is $H_2PO_4^-$ in our case at pH 2.0.

   x is equal to 9.8 mM of $H_2PO_4^-$.

   Table 5-3 shows the concentrations of dihydrogen phosphate after addition of various amounts of phosphoric acid to 1 l of water. It can be seen that a significant amount of phosphoric acid needs to be added to obtain increasing concentrations of $H_2PO_4^-$ once the pH is below the pK_a of phosphoric acid (2.1).

Table 5-3. Dihydrogen phosphate ion concentration and pH at various concentrations of phosphoric acid.

mL acid added	Conc. H_3PO_4 [mM] added	Conc. $H_2PO_4^-$ [mM] corrected for ionization	pH
0.3	4.3	3.1	2.5
0.5	7.3	4.6	2.4
1.5	21.9	9.8	2.0
2	29.2	11.7	1.9
3	43.8	15.1	1.8
4	58.5	17.9	1.7
7	102.3	24.8	1.6
14	204.6	36.5	1.4

If we have a certain concentration of phosphate buffer and then adjust with phosphoric acid how do we calculate the amount of $H_2PO_4^-$ present?

Buffered solution requires a greater amount of acid for the adjustment to a certain pH compared to the non-buffered solution. The amount of acid a buffer can absorb without changing its pH significantly depends on the initial concentration of buffer and amount of acid added.

Let us take for example 1 l of a 25 mM sodium dihydrogen phosphate solution. Then consider that the adjustment of the pH was performed with 1.5 ml phosphoric acid and gave a resultant pH of 2.33.

1. First determine the dominant equilibrium in solution at pH 2.33. Remember, when a salt such as NaH_2PO_4 dissolves, it dissociates into its ions.

$$H_3PO_4 \Leftrightarrow H^+ + H_2PO_4^-, K_{a1} = \frac{[H^+][H_2PO_4]}{[H_3PO_4]}, pK_{a1} = 2.1 \tag{5.1}$$

$$H_2PO_4^- \Leftrightarrow H + HPO_4^{-2}, K_{a2} = [H^+][HPO_4^{-2}], pK_{a2} = 7.2 \tag{5.2}$$

There are two equilibria to be considered. However, if the pH of the resultant solution is more than 2 pH units lower than the pK_{a2} of this polyprotic acid then only the first equilibrium needs to be considered. The first equilibrium is the dominant equilibrium and the H_3PO_4 dissociation reaction involves both the $H_2PO_4^-$ and H_3PO_4 species. The equilibrium expression is denoted by K_{a1} above.

2. Now the concentration of the phosphoric acid added must be calculated. The concentration of phosphoric acid can be calculated as in the prior example using the following equation:

$$\frac{V_{acid} \cdot C_{acid} \cdot d_{acid}}{MW \cdot V_{aq}} = C \qquad \frac{1.5\,\text{ml} \cdot 0.85 \cdot 1.685 \frac{\text{g}}{\text{ml}}}{98 \frac{\text{g}}{\text{mol}} \cdot 1\,\text{l}} = 0.0219\ \text{M}.$$

The initial concentration of H_3PO_4 is 0.0219 M.
The concentration of $H_2PO_4^-$ is equivalent to the buffer concentration since it is assumed that it will be predominately as $H_2PO_4^-$ below, two pH units of its pK_a. Therefore, the initial concentration of $H_2PO_4^-$ is 0.025 M.

4. Consider initial and equilibrium conditions.
Initial conditions are concentrations of the species before any acid dissociation actually occurs.

Generally, $[H_3PO_4] = 0.0219$ M, $[H_2PO_4^-] = 0.025$ M and $[H^+] \simeq 0$ M
Next, consider the change, which is denoted as x, as the change required to reach equilibrium. Doing so we get the following.

$$H_3PO_4 \Leftrightarrow H^+ + H_2PO_4^-$$

Initial	0.0219	0	0.025
Equilibrium	$0.0219 - x$	x	$0.025 + x$

5. Substitute the concentrations at equilibrium conditions into the equilibrium expression and we obtain

$$\frac{x \cdot (0.025 + x)}{(0.0219 - x)} = K_a \longrightarrow \frac{x \cdot (0.025 + x)}{(0.0219 - x)} = 7.94\text{E} - 03$$

Now let us use the quadratic equation and solve for x.

$$x_2 + (0.025 + 7.94\text{E} - 03) \cdot x - (7.94\text{E} - 03 \cdot 0.0219) = 0$$

x is the concentration of H^+ which is 0.0047 M. The concentration of A^-, which is $H_2PO_4^-$ in our case, is $0.025 + x$ and the resultant concentration is 0.0297 M or 29.7 mM. The concentration of HA, which is H_3PO_4, is 0.0219 M – x.
Table 5-4 shows the final concentration of dihydrogen phosphate after varying the initial buffer concentrations. The same amount of phosphoric acid was added to 1 l of the buffers. As the buffer strength increases the system becomes more resistant to pH change. The pH measured of a high concentration buffer (>50 mM) may give a value that deviates from the theoretical pH. This may be due to large differences in the ionic strength of the buffer compared to the standards used to standardize the pH meter. If high concentrations of buffer are needed then the pH meter used should be calibrated with standards of comparable ionic strength.

Table 5-4. Calculated concentrations of sodium dihydrogen phosphate and theoretical pH.

Volume of phosphoric acid (ml) added to 1l	**Initial concentration of sodium dihydrogen phosphate (mM)**	**Final concentration of $H_2PO_4^-$ (mM)**	**Theoretical pH**
1.5	5	13.2	2.08
1.5	10	17.0	2.16
1.5	25	29.7	2.33
1.5	0	52.9	2.53
1.5	100	101.6	2.79

There are several important points to be considered when employing phosphate buffers. The highest purity of buffer salt should be obtained since trace impurities may accumulate in the column and then interfere with subsequent analysis. The phosphate buffer salt is hygroscopic, and overexposure to the atmosphere will lead to significant changes in the composition of the salt. Therefore incorrect concentrations of buffer may be produced since the salt is not solely sodium phosphate. Also, if high concentrations of buffer are needed, a test should be done to see if the buffer salt will precipitate at the organic concentration used for the HPLC. This can be done in a test tube. For example if a 50 mM sodium phosphate buffer is used and the HPLC method requires 50 % organic, then make a solution of 5 ml of 50 mM sodium phosphate buffer and 5 ml of organic solvent in a test tube and shake. If a precipitate is formed then the organic concentration is too high for this concentration of buffer. Either the concentration of the buffer or the organic content should be decreased and the test tube experiment repeated until no precipitate is observed.

Is there a way to maintain the pH and simply adjust the concentration of the counteranion of the acidic modifier?

Yes, you can adjust the pH of the mobile phase by addition of an acid (preferably with low pK_a) and then increase counteranion concentration by adding its salt. The retention increase will be solely due to the increase of the chaotropic counteranion. This approach may be needed to fine tune a method. For example, if a mixture of acids and bases is not optimally resolved at a certain pH then the addition of perchlorate anion will increase the retention of only the protonated basic compounds without affecting the retention of the acidic compound or other basic compounds that are not fully protonated at this pH. In order to calculate the total concentration of perchlorate anions present, the concentration of perchlorate anion from the addition of perchloric acid and sodium perchlorate must be known.

Assume that the pH of the mobile phase is 2.0. This was obtained by adding 1.0 ml perchloric acid to 1 l HPLC Grade Water. The resultant pH was 2.0 and the concentration of the perchlorate anion was 14 mM (see Table 5-1). Now we want to increase the concentration to 50 mM perchlorate anion without changing the pH. Therefore, we need an additional 36 mM of perchlorate anion. The sodium perchlorate in water is completely dissociated and therefore 1 mol of sodium perchlorate is equal to 1 mol of perchlorate anion. Since we know that 36 mM sodium perchlorate is equal to 36 mM perchlorate anion we now have to calculate the number of grams of sodium perchlorate (MW 122 g/mol) to obtain the desired concentration.

$$0.036 \frac{\text{mol NaClO}_4}{\text{l}} \cdot 122 \frac{\text{g NaClO}_4}{\text{mol NaClO}_4} = 4.4 \text{ g NaClO}_4$$

4.4 g sodium perchlorate added to 1 l water will give a concentration of perchlorate anion of 36 mM. Hence, 36 mM perchlorate from sodium perchlorate + 14 mM perchlorate from perchloric acid equals a total concentration of 50 mM perchlorate anion. The final pH of this mobile phase will be 2.0 since the addition of salt will not effect the pH. We are not adding more H^+ (protons) to the mobile phase therefore the pH will not change.

The sodium perchlorate salt was added, but what other salts that contain chaotropic counteranions could I use?

Any alkali salts of PF_6^-, BF_4^- and $CF_3CO_2^-$, may be used. These all have a high charge delocalization and low degree of hydration.

This chaotropic effect is predominant under highly aqueous conditions, but is it significant where the organic content is higher?

An increase in the organic content in the mobile phase suppresses the effect of the anionic chaotrope on the basic analyte retention. The plot of k vs concentration of

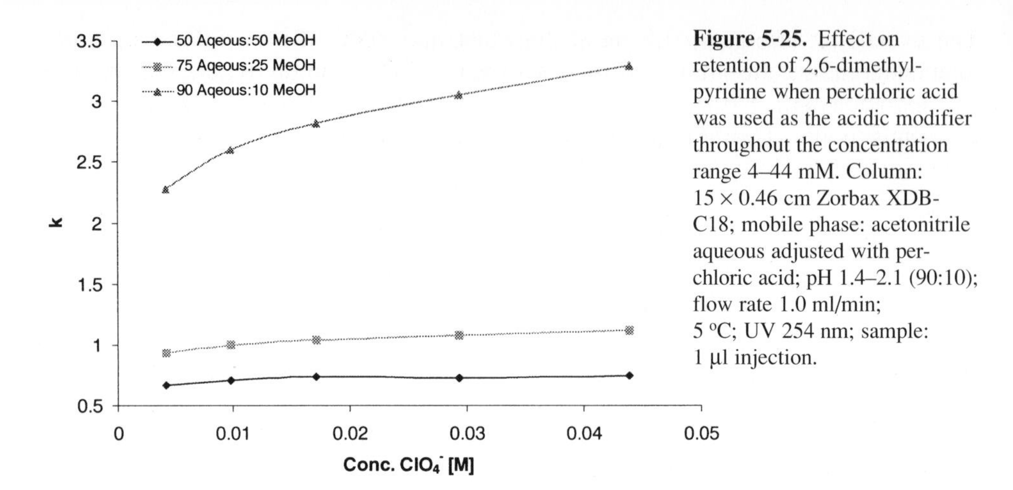

Figure 5-25. Effect on retention of 2,6-dimethyl-pyridine when perchloric acid was used as the acidic modifier throughout the concentration range 4–44 mM. Column: 15 × 0.46 cm Zorbax XDB-C18; mobile phase: acetonitrile aqueous adjusted with perchloric acid; pH 1.4–2.1 (90:10); flow rate 1.0 ml/min; 5 °C; UV 254 nm; sample: 1 μl injection.

perchlorate (Fig. 5-25 below) demonstrates that the greatest effects are seen in the 90 % aqueous (adjusted with perchloric acid): 10 % organic (methanol) eluent.

Therefore, it is recommended to work at more highly aqueous conditions in order to obtain a more significant chaotropic effect. The more water the mobile phase has the higher is the degree of its structurization and the higher the degree of the basic analyte solvation. The appearance of the chaotropic counteranions in these highly aqueous mobile phases causes a significant destructive effect on the analyte solvation shell. On the other hand, in highly organic mobile phases the analyte solvation is minimal, so there are less water molecules for counteranions to disrupt.

What is the influence of temperature on the chaotropic effect?

The overall chaotropic effect within a given concentration range is not similar at different temperatures. A plot of k vs counteranion concentration is given in Fig. 5-26.

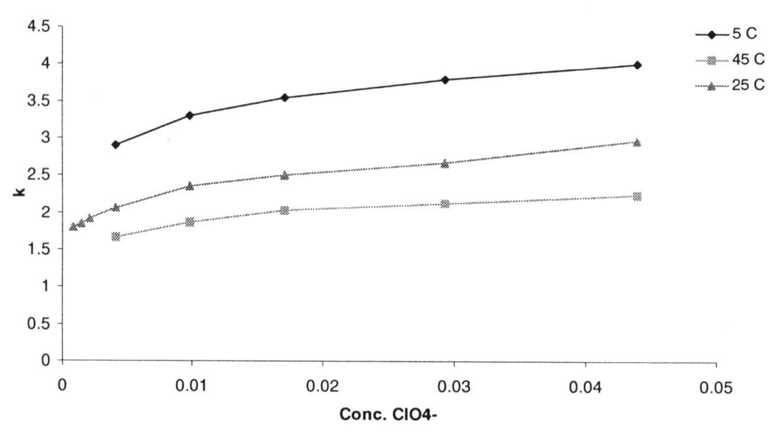

Figure 5-26. Effect on retention of aniline when perchloric acid was used as the acidic modifier throughout concentration region 1–44 mM. Column: 15 × 0.46 cm Zorbax XDB-C18; mobile phase: methanol-aqueous adjusted with perchloric acid; pH 1.4–2.1 (90:10); flow rate 0.6 ml/min; 5 °C, 25 °C, 45 °C; UV 254 nm; sample: 1 μl injection.

The overall retention factor increase for the basic compound increases within the temperature range from 5–45 °C. It is speculated that at the lower temperatures the solvation around the basic analytes is more ordered and that small changes in the environment would lead to significant changes in the retention of the basic compound with increased concentration of the counteranion.

What is the optimal temperature at which method development should be carried out?

The temperature of the analysis may at times be critical for the separation. Temperature generally has a small effect on band spacing in reversed-phase chromatography for neutral samples. However, this may not be so for ionic samples since different retention processes can be affected differently by a change in temperature. This may be especially true when analyzing a basic compound near its pK_a since the temperature effect on the retention of the ionized molecules may be different from the temperature effect on unionized molecules of the same compound. Temperature also affects the analyte ionization equilibrium, so this will cause a shift in the analyte pK_a. Both these effects together may significantly influence the chromatographic selectivity. Therefore, the column temperature should be properly thermostated preferably with a water jacket.

Is there a way to calculate the pK_a of my compound within a mixture if only a limited amount is available.

Yes, if the compound of interest is well resolved from other components in your mixture the pK_a may be determined.

- First, to absolutely confirm the structure of the compound of interest LC-MS should be done if pure sample is available do spiking experiments to confirm retention.
- In order to determine the pK_a, prepare the aqueous portion of the mobile phase with a pH of 8, 7, 6, 5, 4, 3, and 2. (You do not need a high buffer concentration.)
- Obtain the retention factors of the compound at each mobile phase pH. Remember to let the column equilibrate when changing the pH of the mobile phase.
- Plot k vs pH. Figure 5-27 shows an example of this plot for various isomeric dimethylpyridines.

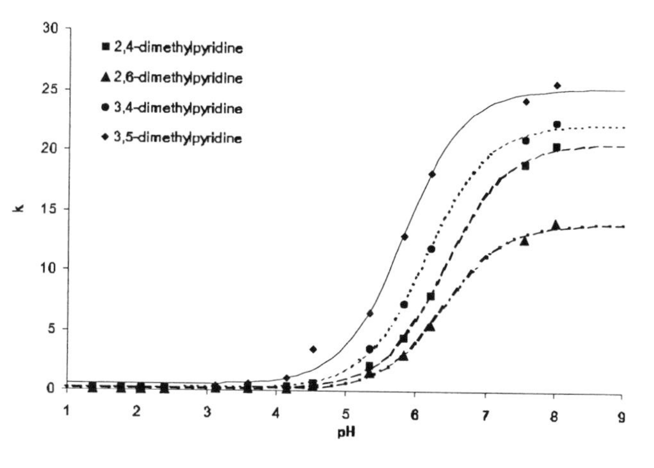

Figure 5-27. Isomeric series of dimethylpyridines. Column: 15 × 0.46 cm Zorbax XDB-C18; mobile phase: acetonitrile-10 mM sodium dihydrogen phosphate buffer adjusted with phosphoric acid; pH 1.3–8.6 (90:10); flow rate 1.0 ml/min; 25 °C; UV 254 nm; sample: 1 µl injection.

- This method is preferable to titration when only impure mixtures or small sample sizes are available.
- From Eq. 5.1 we can obtain the pK_a at the inflection point from the plot of retention factor vs pH in the particular binary eluent system.

This approach actually offers the chromatographer the advantage of obtaining an estimated pK_a for a compound which is not available in the isolated pure form.

We should mention that the actual pH of the organic-buffer mixture is not the same as the pH of the buffer alone. Below is a comparison of pK_a values for a series of different basic components obtained from the literature (Table 5-5, column 2) and pK_a experimental values measured using the technique described above with three different acidic modifiers.

Table 5-5. Determination of pK_a using reversed-phase HPLC.

Compound	Literature pK_a 25 °C in Water	H_3PO_4 exp pK_a 25 °C	TFA exp pK_a 25 °C	$HClO_4$ exp pK_a 25 °C
pyridine	5.17	4.85	4.75	5.03
2-ethylpyridine	5.89	5.62	5.56	5.62
3-ethylpyridine	5.80 (20 °C)	5.42	5.3	5.43
4-ethylpyridine	5.87	5.8	5.66	5.8
2,4-dimethylpyridine	6.74	6.42	6.27	6.31
2,6-dimethylpyridine	6.71	6.41	6.27	6.27
3,4-dimethylpyridine	6.47	6.2	6.02	6.08
3,6-dimethylpyridine	6.09	5.82	5.72	5.76
aniline	4.6	4.08	4.13	4.2
N-methylaniline	4.85	4.4	4.44	4.64

All literature pK_a values were determined at 25 °C in water unless otherwise noted [13].

All experimental pK_a values were determined in a 90 % aqueous buffer containing 10 mM sodium phosphate adjusted with perchloric, trifluoroacetic or phosphoric acid and 10 % acetonitrile at 25 °C.

As one can see, there is a difference between the compound pK_a in buffer alone and in the solvent which contains 10 % of acetonitrile. The observed difference is about 0.3 pH units for all measured compounds.

The pK_a value is actually a pH at which the concentrations of ionized and unionized forms of the analyte are equal. The actual pH of the mixed eluent containing organic is higher than that of pure buffer. Thus, the analyte retention will be dependent on these higher pH values. However, the retention factors are associated with the pH of the buffer before addition of organic. This leads to a shift of pH dependence of the analyte retention to lower values. Therefore, the pK_a value calculated from this dependence will be lower than the true value.

The effect of acetonitrile concentration on the pH of buffered water/acetonitrile solutions has been studied by J. Barbosa and V. Sanz-Nebot [14]. They showed that each 10 % increase of acetonitrile content in water causes a pH increase of approximately 0.3 units. Their experimental data measured for different buffers are shown in Fig. 5-28.

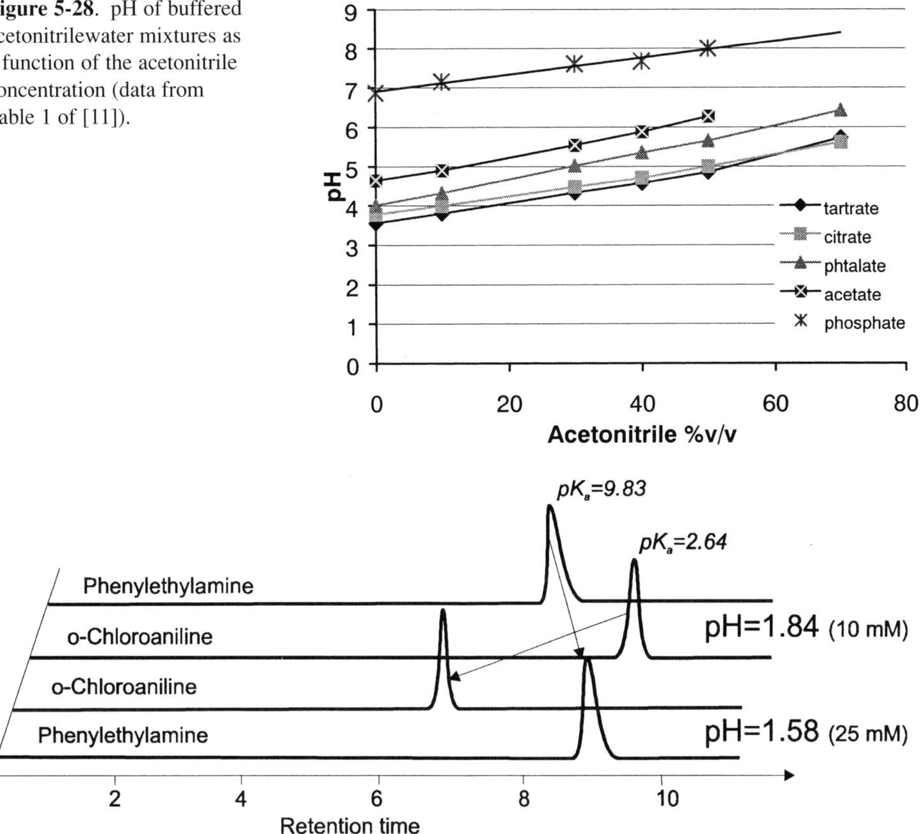

Figure 5-28. pH of buffered acetonitrilewater mixtures as a function of the acetonitrile concentration (data from Table 1 of [11]).

Figure 5-29. Influences of change in concentration on retention of basic analytes.

The slopes of all the lines in the above figure are almost equal. As one can see, the addition of 50 ml of acetonitrile to 50 ml of acetate buffer will increase its pH by more than 1.5 units.

We should also mention that the actual pH measurements of the waterorganic mixtures could only be done using special techniques (potentiometry), any pH meter will show you results which could be better attributed to the pH of water in, say, lake Michigan 30 years ago.

How significant will the retention shift be if a change of the counteranion concentration occurs in a low pH region?

As can be seen in Fig. 5-29, the retention of the two basic compounds *o*-chloroaniline (pK_a 2.64) and phenylethylamine (pK_a 9.83) were affected in different ways.

The retention of phenylethylamine increased by approximately 1.5 min with an increase in counteranion concentration from 10 mM (pH 1.84) to 25 mM (pH 1.58). At pH 1.84, phenyethylamine is fully protonated. As the counteranion concentration was increased, the ion association occurred, causing decrease of solvation of the phenylethylamine analyte and making it more hydrophobic. The phenyethylamine retention change was governed by the chaotropic effect (counteranion concentration). On the other hand, the retention of *o*-chloroaniline decreases with the pH decrease from 1.84 to 1.58. Since the p*K*a of *o*-chloroaniline is 2.64, it is not fully protonated in this pH region and its retention change was governed by ionization. As the pH of the mobile phase decreased, this compound was becoming more ionized and thus more polar, which causes its retention decrease. The change in the analyte retention due to its ionization is much more significant than the chaotropic effect.

What other advantages are obtained from using this chaotropic approach?

The advantage of using a combination of low pH and a higher perchlorate concentration is that at these low pHs interactions with the silanols or secondary equilibria effects will be non-existent.

This approach may be used as an alternative method in order to obtain better peak shapes and increased retention of basic components. Figure 5-30 shows an example of the increase in retention of aniline when perchloric acid was used as the acidic modifier.

As can be seen from Fig. 5-30, the retention factor decreases together with the pH of the mobile phase, until pH 2.6 is reached. Then the retention factor starts to increase with the increase of the concentration of the perchlorate anion (decrease in pH). This is due to the chaotropic effect. At pHs close to the pK_a of aniline (4.6), the peak shape is broad and severe fronting is observed. The increase of the perchlorate concentration at low pHs gave retention factors comparable to those at higher pH values.

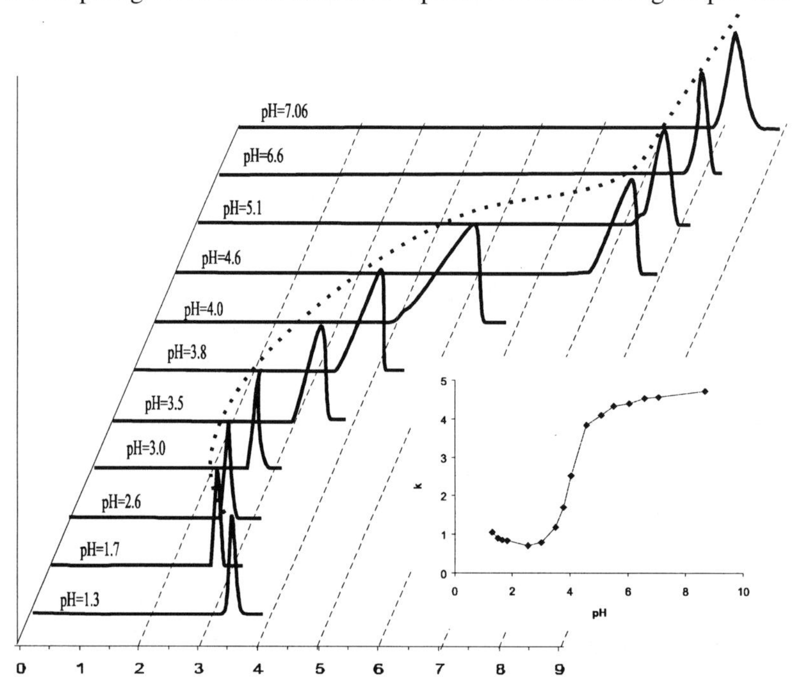

Figure 5-30. Effect on retention of aniline when perchloric acid was used as the acidic modifier through the pH range 1.3–7.1. Column: 15 × 0.46 cm Zorbax XDB-C18; mobile phase: acetonitrile10 mM sodium dihydrogen phosphate buffer adjusted with perchloric acid; pH 1.38.6 (90:10); flow rate 1.0 ml/min; 25 °C; UV 254 nm; sample: 1 µl injection.

5.4 Concluding remarks

We have discussed so far the general approach to method development for the separation of ionizable compounds, emphasizing modification of the mobile phase. Despite the fact that this approach was designed for conventional reversed-phase HPLC columns and a standard HPLC set-up, there are some important requirements for columns, analyte solutions, and reagents.

Column pH stability is an important factor which has to be considered before the method development starts. Silica-based reversed-phase columns could degrade at high mobile phase pH of the mobile phase is high due to dissolution of the silica or at low pH due to hydrolysis of alkylsilane bonds. The most important parameter affecting these processes is the bonding density of the stationary phase. The higher the bonding density the better the shielding of the underlying silica surface. This prevents possible attack of that surface and its dissolution or hydrolisation.

Most of the recently developed commercial C18-type phases (monomeric) based on high purity silica have shown long-term stability over a wide pH range. As examples we could mention several phases we have used at extreme pHs and did not notice practically any degradation over several months of use.

- Prodigy-ODS2 (Phenomenex, Torrance, CA)
- Zorbax-Eclipse XDB-C18 (Hewlett-Packard, Little Falls, MD)
- Symmetry-C18 (Waters, MA)

Highest possible purity of used reagents is an absolute requirement for the mobile phase preparation. You are pumping liters of the mobile phase through the column and if there are any impurities in the salt or acid used for the pH adjustment, then this impurity will be trapped on the adsorbent surface and will gradually change its selectivity. In most of the cases so called column degradation is not real column degradation but a severe contamination of the stationary phase with the impurities from the mobile phase.

It is possible to clean this highly contaminated column by forcing a series of the solvent fronts through it. The best way is pump pure water at ~ 1 ml/min for 2 min then proceed to pure acetonitrile for another 2 min and repeat this cycle several times. Sharp solvent front passing through the column causes a significant disturbance in the stationary phase forcing the release of most of the trapped contaminants. The cleaning effect of that single front is equivalent to hours of washing with pure acetonitrile.

The separation of a complex mixture of different ionizable and nonionizable organic components is always a challenge, and the development of the rugged separation method is an adventure.

Two different ways exists to approach this challenge.

One is to find a column with specific selectivity for particular components in your mixture. There are thousands of different columns on the market and the selection of one, which will allow a simple separation of your mixture, is more a black magic than a science.

The other approach is based on the use of current systematic (although limited) knowledge of the reversed-phase mechanism in combination with the principles of the analyte behavior in the mobile phase. This offers some flexibility in the controlling the selectivity of your method and certain predictability of components retention. Although some black magic may still be needed there also.

5.5 References to Chapter 5

[1] C.Horvath, B.A.Preiss, and S.R.Lipsky, *Anal. Chem.* 39, (1967) 1422.
[2] C.Horvath and S.R.Lipsky, *J. Chromatogr. Sci* **7**, (1969) 109.
[3] Horvath, Csaba, High Performance Liquid Chromatography:Advances and Perspectives, Vol.2, Academic Press, New York, 1980.
[4] L.R.Snyder and J.J.Kirkland, Introduction to Modern Liquid Chromatography,Wiley Interscience, New York, 1974, Chapter 9.
[5] J.T.Eleveld, H.A.Claessens, J.L.Ammerdorffer, A.M.van Herk, and C.A.Cramers, *J. Chromatogr.* A, **677,** (1994) 211.
[6] J.A.Lewis, D.C.Lommen, W.D.Raddatz, J.W.Dolan, and L.R.Snyder, *J. Chromatogr.* **592,** (1992) 183.
[7] Qian-Hong Wan, Martyn C.Davies, P. Nicholas Shaw, and David A. Barrett, *Anal. Chem.*, **68**, (1996) 437-446.
[8] P.J.Shoenmakers and R.Tijssen, *J. Chromatogr.* A, **656,** (1993) 557.
[9] C.Horvath, W.Melander, and I.Molnar, *Anal. Chem.*, **49,** (1) (1977) 142.
[10] J.J.Kirkland, M.A.vanStraten, and H.A.Claessens, *J. Chromatogr.* A., **797,** (1998) 111-120.
[11] A.Ishikawa and T.Shibata, *J. Liq. Chromatogr.***16,** (1993) 859.
[12] Y.Hatefi and W.G.Hanstein, Proc. Natl. Acad. Sci. U.S.A **62,** (1969) 1129-1136.
[13] Langés Handbook of Chemistry, 4th edn.
[14] J.Barbosa and V.Sanz-Nebot, Preferential solvation in Acetonitrile Water Mixtures, *J. Chem. Soc. Faraday Trans.* **90,** (1994) 3287-3292.

6. Appendix

6.1 Some chromatographic and related abbreviations (selection)

Almost monthly you can find new abbreviations in the literature, usually referring to new separation mechanisms or new procedures. You often lose track of these, so I began to write down the most common ones out of a necessity. I hope this list will also be helpful to you.

Abbreviation	Term
AC	Affinity Chromatography
APCI	Atmospheric Pressure Chemical Ionization
AP-ES	Atmospheric Pressure Electro Spray
CC	Computational Chromatography Cocurrent Chromatography
CCC	Counter-Current Concentration or Continuous Countercurrent Chromatography
CCCC	Continuous Cross-Current Chromatography
CCE	Counter Current Electroconcentration
CE	Capillary Electrophoresis
(C)EC	Capillary Electrochromatography
CEPA	Capillary Electrophoresis with Polyacrylamide
CGC	Chiral Gas Chromatography
CGE	Capillary Gel Electrophoresis
CIA	Capillary Ion Analysis
CIEF	Capillary Isoelectric Focusing
CIS MS	Coordination Ion Spray Mass Spectrometry
CLC	Capillary Liquid Chromatography Chiral Liquid Chromatography
CPC	Centrifugal Partition Chromatography
CZE	Capillary Zone Electrophoresis
FIA	Flow Injection Analysis
FPLC	Fast Protein Liquid Chromatography
FZCE	Free Zone Capillary Electrophoresis
GC	Gas Chromatography
GFC	Gel Filtration Chromatography
GLC	Gas-Liquid Chromatography
GPC	Gel Permeation Chromatography
GSC	Gas-Solid Chromatography
HATP	High Aqueous Tolerant Phases
HDC	Hydrodynamic Chromatography
HIC	Hydrophobic Interaction Chromatography
HILIC	Hydrophilic Liquid Chromatography
HPCE	High Performance Capillary Electrophoresis
HPCEC	High Performance Capillary Electro Chromatography
HPEC	High Performance Electro Chromatography
HPMC	High Performance Membrane Chromatography
HPOTLC	High Performance Open Tube Liquid Chromatography

HPST	High Performance Separation Techniques
HPTLC	High Performance Thin Liquid Chromatography
HSLC	High Speed Liquid Chromatography
HTOTLC	High Temperature Open Tube Liquid Chromatography
HTP	High Throughput Purification
HTA	High Throughput Analysis
HTS	High Throughput Screening
I(L)C	Ion (Liquid) Chromatography
IEC	Ion Exchange Chromatography Ion Exclusion Chromatography
IMAC	Immobilized Metal Affinity Chromatography
IPC	Ion Pair Chromatography
ISRC	Internal Surface Reversed Phase
LC	Liquid Chromatography
LC-MS/LC-ESI-MS	Liquid Chromatography Mass Spectrometry/Liquid Chromatography Electro Spray Interface Mass Spectrometry
LEC	Ligand Exchange Chromatography
LLC	Liquid Liquid Chromatography
LSC	Liquid Solid Chromatography
μ-TAS	micro-Total Analytical System
MC	Mass Chromatography
MCIC	Metal Chelate Ion Chromatography
MDGC	Multi Dimensional Gas Chromatography
MEC	Micellar Electrokinetic Chromatography
MECC	Micellar Electrokinetic Capillary Chromatography
MICA	Molecularly Imprinted Chromatography Analysis
MIP	Molecularly Imprinted Phases
MPLC	Middle Pressure Liquid Chromatography
NAC	Non Aqueous Chromatography
NACE	Non Aqueous Capillary Electrophoresis
NARP	Non Aqueous Reversed Phase
NPC	Normal Phase Chromatography
OCEC	Open (tube) Capillary Electro Chromatography
OTEC	Open Tube Electro Chromatography
OTLC	Open Tube Liquid Chromatography
PIC	Paired Ion Chromatography
PS-DVB	Poly-Styrene-Divinyl Benzene
RP-IPC	Reversed-Phase Ion Pair Chromatography
RPC	Reversed-Phase Chromatography
SDS-PAGE	Sodium Dodecyl Sulfate Polyacryl Amide Gel Electrophoresis
SEC	Size Exclusion Chromatography
SFC	Supercritical Fluid Chromatography
SFE	Supercritical Fluid Extraction
SIM	Selected Ion Monitoring

SPE	Solid Phase Extraction
SPME	Solid Phase Micro Extraction
UHC	Ultra Highspeed Chromatography

6.2 IUPAC recommendations for symbols in chromatography (a selection)

IUPAC (International Union of Pure and Applied Chemistry) published in 1993 in Pure and Applied Chemistry, Vol. 65, No. 4, pp. 819872 recommendations for „Nomenclature for Chromatography". The 53 page paper is reprinted in the „ChromBook" from the Merck company, Darmstadt. Below, the most important parameters with their symbols are listed.

Parameter	**Symbol**
Separation factor	α
Selectivity factor (up to 1993)	α
Area	a
Diameter	d_c
Diffusion coefficient	d
Porosity	ε
Flow rate (volumetric)	f
Plate height	h
Viscosity	η
Equilibrium (distribution) constant	k
Rate constant	k
Retention factor	k
Capacity factor	k'
Length	l
Plate number	n
Density	ρ
Pressure	p
Pressure (relative)	p
Radius	r
Temperature (absolute)	t
Time	t
Retention time	t_r
Velocity (linear)	u
Volume	v
Retention volume	v_r
Mass (weight)	w
Peak width	w

6.3 Solvent mixtures of equal elution strength for reversed phase chromatography (according to L. Snyder)

Using the following Table, you can make different mixtures of mobile phases to give the same elution strength. For example, you can expect almost the same retention time from the mixtures of 50/50 methanol/water, 40/60 acetonitrile/water and 30/70 THF/water – if there are no specific interactions. In addition, you can use the following rule of thumb: a 10 % change of the organic part of the mobile phase changes the *k* values and the retention times by a factor of 2 to 3. With this, you can roughly predict the results of your optimization experiments.

Methanol/water	Acetonitrile/ water	THF/water	*k*
0	0	0	100
10	6	4	40
20	14	10	16
30	22	17	6
40	32	23	2.5
50	40	30	1
60	50	37	0.4
70	60	45	0.2
80	73	53	0.06
90	86	63	0.03
100	100	72	0.01

6.4 UV absorption bands and molar extinction coefficients of some typical chromophores

In most cases the UV behavior of the components to be separated is known. If this is not so in your case and you furthermore do not have a Diode Array Detector at your disposal, the following table will help you. If the structure of the components to be separated is known, you can read off the suitable detection wave length, dependent on the chromophore group. For reasons of clarity only the most important chromophores were considered.

Name of chromophore	Structure	$\lambda_{max,1}$	$A_{max,1}$	$\lambda_{max,2}$	$A_{max,2}$	$\lambda_{max,3}$	$A_{max,3}$
Alkine (Ethine)	-C≡C-	175-180	6000				
Aldehyde	-CHO	210	groß	280-300	11-18		
Amine	$-NH_2$	195	2800				
Anthracene	Φ-Φ-Φ	252	199000	375	7900		
Azido	>C=N-	190	5000				
Azo	-N=N-	285-400	3-25				
Benzene	Φ	184	46700	202	6900	255	170
Bromide	-Br	208	300				
Carboxyl	-COOH	200-210	50-70				
Diphenyl	Φ-Φ-	246	20000				
Disulfide	-S-S-	194	5500	255	400		
Ester	-COOR	205	50				
Ether	-O-	185	1000				
Ethylene	-C=C-	190	800				
Iodide	-I	260	400				
Isoquinoline	Φ-Φ((2)N)	21	80000	266	4000	317	3500
Ketone	>C=O	195	1000	270-285	18-30		
Naphthalene	Φ-Φ	220	112000	275	5600	312	175

6.5 List of tables

6.6 HPLC textbooks

There are many books about HPLC. A short list of the books I like most is given below (in order of year of publication):

L. R. Snyder, I. I. Kirkland:	Introduction to Modern Liquid Chromatography Wiley-Interscience, 1988 (Old, but still good for understanding the topic, a "classic")
J. W. Dolan, L. R. Snyder:	Troubleshooting LC Systems Humana Press, 1989 (Old too, but still valid and very helpful for every-day problems.)
V. Meyer	Pitfalls and errors of HPLC in pictures (Small and good, easy to read)
U. Neue:	HPLC columns Wiley-Interscience, 1997 (A good overview and explanation of stationary phases not suitable for absolute beginners.)
B. Bidlingmeyer	Practical HPLC Methodology and Applications
Snyder, Glajch, Kirkland	Practical HPLC Method development (Both the last ones are suitable for careful study of a method development and optimization.)
P. Sadek	Troubleshooting HPLC Systems, a Bench Manual (general hints, maintenance and troubleshooting of the HPLC equipment).
Dictionaries:	
H. P. Angel:	Dictionary of Chromatography (English, German, French, Russian), Alfred Hüthig Verlag, 1984, ISBN 3-7785-0926-8

6.7 Trends in HPLC

In specialized journals and laboratory literature, you can find excellent overviews describing new developments and trends revealed in important symposia, meetings and conventions and trade shows, for example:

International Symposium on Chromatography
International Symposium of High Performance Liquid Phase Separations and Related Techniques
International Symposium on Column Liquid Chromatography
Pitcon
Analytica

Please find some of the most important trends and developments in a short form below.

General Trends

HPLC is a mature, very flexible and, given the complexity, quite robust method.

- The instrumentation is being developed towards simpler designs of the mechanics, optics and electronics, resulting in simplified maintenance. And the instruments are becoming smaller and easier to handle.
- The software is extremely powerful and extensive, sometimes even too extensive for routine analysis. It is also often complicated and cumbersome.
- Automation in HPLC is already at a high level. Sample preparation is still a major issue and there is a considerable demand to catch up with the backlog. Sample preparation and injection co-develop into one step in the analysis, often automated as well.
- The single modules (pump, autosampler, detector) often contain diagnostic functions to give information about their history and current status, such as volume of mobile phase delivered since the last maintenance check, lamp energy, air in syringe, etc.
- Miniaturization of the column is now slowly appearing on the agenda, and at many places is at least a discussion subject. An important hurdle to be overcome when promoting a shift towards lower diameter and length of the columns is certainly the immense effort necessary in strongly regulated laboratories to admit and revalidate new methods.
- "Coupling" is the focus subject today and surely for the coming years.
- The big development steps are predominantly in the stationary phase, electrochromatography and naturally in the coupling techniques; then in detection systems. A really big development step forward in detection can only be conceived with laser technology and miniaturization.
- To put it briefly, we have at moment in the area of liquid separation techniques a coexistence of HPLC and the related techniques CLC, CE and CEC. Classic HPLC is the dominant technique in real-life laboratories, with increasing emphasis on miniaturization (length up to 30–60 mm, I.D. up to 2–3 mm, dp up to 2–3 µm) while CLC, CE and CEC are the champions in the absolute research environment and in symposia.

Let us now have a closer look at the individual HPLC modules

Pump	Today's pumps provide excellent flow reproducibility with a V_C of 0.5 % and often even lower. Isocratic pumps find fewer and fewer buyers; users obviously like to be able to run gradients.
Gradient	V_C with modern instrumentation is better than 1 %, often around 0.5 %. Users seem to slightly prefer low-pressure gradient mixers over high pressure gradient mixers. Degassing units are often built-in. Many instruments are capable of working with miniaturized columns.
Injector	
(a) Manual	• Loops in the nl-range for μ-bore applications or capillary LC (see below) are commercially available • Manual valves made from inert materials have been commercially available for a couple of years, but the demand is rather low
(b) Autosampler	Simple steps in sample preparation such as mixing, pretexing, filtration, dilution, making aliquots, etc. are already implemented in many instruments. Some manufacturers even allow extremely small injection volumes for μ-bore techniques
Column	You should always be aware of the Pareto-principle: in a US $ 50 000 instrument this "tube", costing US $ 300–400, is still the most important part ..
(a) Design	• The typical column today is 125–200 mm in length with an I.D. of 4–4.6 mm and filled with 5 μm spherical particles • The breakthrough for the 3 μm particles and 30–60 mm length is coming soon • For method development, 3 and even 2 mm ID columns and even more 60 mm length columns are used more and more. Capillaries (< 250 mm) are a dominant subject at HPLC symposia however only there. Miniaturization of the column is a good example of the time delay between technical feasibility and routine application: using a 3 mm column instead of a 4.6 mm column, the mobile phase savings amount of 60 % in addition to an increased peak height of 2, assuming the same packing quality. A change from 4 mm to 3 mm results in mobile phase savings of 45 % and a peak height increase of 1.5. Although many instruments were capable of handling small bore columns and their advantages wildly publicized, miniaturization became a subject for everybody only in the late 1990s.

(b) Chemistry

Silica gel from most manufacturers is, after a thorough acid wash with HCl or H_2SO_4 at low temperature, for the most part "clean" and metal free. New approaches are silica gel made from alkoxysilanes to get a pure material (e.g. Symmetry, Purospher) and monolithic media (SilicaROD).

C_{18} is the undisputed favorite of all stationary phases. Next in the top ten list is, after a huge gap, C_8, C_4, CN, NH_2, SiO_2, Phenyl, Diol. Ion exchange and GPC columns have a constant and true customer base. Affinity media have in biochemistry similar true fans.

The current development in reversed-phase chromatography goes in the direction of sterically or chemically protected phases (shield), phases with hydrophilic endcapping (AQ, EPS) and hybrid phases (XTerra).

The (weak) opposition to silica gel consists of non porous materials, polymer phases, mixed phases (differing surface coverage within one column), titanium dioxide TiO_2, aluminum oxide Al_2O_3, graphite and micro cellulose.

The shooting stars of the last years are chiral phases and pH-stable RP-phases.

Stable polymer phases (e.g. molecularly imprinted gels) and non-porous materials have the best chances for a broader use. However, especially for non-porous materials the high price is an issue.

Detector

UV_{var}, possibly as DAD, is still *the* detector in HPLC. Again, on the top ten list, there is a huge gap before the fluorescence detector, again a gap and then the electrochemical detector, light scattering and FTIR. In carbohydrate, mineral oil and polymer analysis, the RI detector is still widely used.

In coupling HPLC to spectroscopy, LC-MS(MS) coupling is the favorite. LC-FTIR, LC-NMR, LC-ICP and LC-AAS are fighting to at least enter the research laboratories. For difficult analytical purposes LC-NMR-MS(MS) or LC-NMR-MS-FSA (FSA: Flow Scintillation Analyzer) is coming.

Data Evaluation

(a) Integrators

They are becoming a rare species.

(b) PC

You almost do not find an HPLC without a PC any more. The PC handles data evaluation, instrument management, optimization of the separation, validation, statistics, statistical process control. The PC is extremely powerful, but its potential is rarely fully appreciated.

Sometimes, it is simply a question of user-friendly applications. Often small changes in reporting are only achieved with macros or small utility programs.

Other

Optimization programs are useful in an environment where method development and separation optimization are a daily business. They are becoming affordable, more powerful and better, and there is a growing choice available. I think the next step will be the combination of this software with experimental factor analysis tools to optimize the separation according to the results of the latest injection and having the requirements of the user in „mind" (e.g. $R = 2$ for peaks Nos. 5 and 6). Columns and eluent switches will be controlled for an automated optimization procedure overnight.

The importance of HPLC-related techniques, such as capillary -LC (CLC), CE and especially electrochromatography (CEC) will increase. CLC is from an instrumental point of view advanced; CEC fast approaching advanced levels.

High-temperature HPLC (methanol at 120 °C, retention times in the range of seconds) and low temperature HPLC (5 to 10 °C), which can be used to analyze solute configurations, are techniques with practical advantages. However, they will for the time being remain in the research laboratories, like even bolder approaches such as the chip HPLC.

Conclusion, Outlook

Specialization, a phenomenon of our times, also happens in analytical chemistry. The widening gap between routine and research analysis strongly influences the daily work and the environment of an HPLC laboratory.

The different requirements put a stamp on the whole system.

In research, the goal is to solve ever more difficult analytical problems, requiring flexibility in terms of methods, technical instrumentation and the researchers involved. Regulation is (still) rather low. We have to be effective, or in other words to choose the right methods from a toolbox, which is a prerequisite to reacting to a new situation in the optimal way.

In routine analysis, standardized total instrument systems are installed and perfected in respect to robustness. The goal is a maintenance free, cleanly defined procedure and in the end efficiency: the production of as large as possible a number of secured data per time unit. This goal influences the whole system. In the following two Sections, characteristics of HPLC in routine analysis and research are compared.

6.7.1 HPLC in routine analysis

The goal here is efficiency, which means: "Do things in the right way". Then economy and profitableness will increase. The need is simply for robust and dependable instruments with a high degree of automation. Every activity in such an environment has to be well known, so that it can be programmed in advance. The whole system must be based on stability (in statistical control) and robustness; every change leads to a decrease of rentability. So the HPLC method is stored in a PC or a server or comes on a disk or CD-ROM. What to do is well documented. The formalism disappears into thin air. There is anyway a danger that the user becomes more and more an instrument servant. Real potentials of the user (human resources) and often also of the instrument are not used appropriately. The unused potential is often overlooked in a cost-cutting exercise. These developments and their consequences are an important subject, but outside the scope of this book .

Each change – even an improvement – in the system does have some up-front costs. Since the economics are unfavorable in the short term, changes are avoided at almost any price, even if they show the potential to maximize the return on investment in the future. A method, once it has been established, has a long life – independently of its analytical quality. Routine HPLC will face intense competition from spectroscopic methods such as quantitative AAS, UV and especially (FT-)NIRS as well as titrimetry and specific instant test procedures. The importance of HPLC as a routine analysis method will be maintained if applications of specific, simple, almost maintenance-free, self-controlled and inexpensive instruments are available and the probability of user errors decreases to a minimum.

6.7.2 HPLC in a research environment

The goal here is efficacy, which means: "Do the right things". Then effectiveness will increase. In research, constantly changing problems are the daily business. The aim is problem solving with efficacy. HPLC is an important part in the analytical chain, but it is only a part. The more complex the problems, the more pressing are off-line/on-line couplings necessary, because difficult separations can almost not be accomplished without at least one coupling step. The best options for realization are in coupling HPLC with the related gel permeation chromatography, thin chromatography, gas chromatography, planar chromatography, capillary electrochromatography and capillary electrophoresis. But also think of coupling with other techniques such as ELISA/Westernblott or further increase the specificity with chemo- or biosensors.

Process oriented thinking will help here to move into uncharted country to increase peak capacity. In an R&D environment users have to decide in the future more and more in terms of: "Do I really need a separation or perhaps a determination technique?" or in other words: "Which is the most efficient and effective way to get the information I need?" The goal will be:

Maximize the value of the expression $\frac{\text{density of information}}{\text{unit time}}$

The importance of powerful tools such as chemometrics and multivariate analysis for this issue will increase.

HPLC instruments for changing requirements consist of a gradient pumping system with a 6 or 12 port column switching valve, a 3 or 4 eluent switching valve and at least a DAD and an MS(MS), fluorescence or another special detector. The interlinking between chromatography, spectroscopy and computer technology will continue to grow, as has been borne out since the beginning of the 1990s by the introduction of the terms "computography, spectrography and chromatoscopy". Will HPLC pose new challenges as an integrated part of a miniaturized total analysis system, m-TAS, as a **H**igh **P**ico **L**iquid **C**hromatography or a **H**igh **P**erformance **L**iquid **C**hip chromatography?

The fun of HPLC is here to stay.

"... and 100 years after Tswett, at the beginning of the 21st century, it is being realized that miniaturization and coupling ..." The history of chromatography lives on.

Wenn du von allem dem, was diese Blätter füllt,
Mein Leser, nichts des Dankes wert gefunden,
So sei mir wenigstens für das verbunden, was ich zurückbehielt.

Lessing

If you, in everything that fills these pages,
Dear reader, found nothing worth your thanks,
So be at least obliged for that which I withheld.

Lessing

Index